手　绘　北　京　系　列　A Sketch of Beijing

A Bite of Beijing
馋了，风味儿北京

袁小茶　文
Written by
Yuan Xiaocha

刘海乐　译
Translated by
Liu Haile

刘天羽
徐翔宇
张伊聪
绘
Illustrated by
Liu Tianyu
Xu Xiangyu
Zhang Yicong

CIPG　中国画报出版社 CHINA PICTORIAL PRESS　北京市人民政府新闻办公室 Information Office of Beijing Municipality

图书在版编目（CIP）数据

馋了，风味儿北京：汉英对照 / 袁小茶文；刘天羽，徐翔宇，张伊聪绘；刘海乐译. -- 北京：中国画报出版社，2024.6
ISBN 978-7-5146-2214-0

Ⅰ. ①馋… Ⅱ. ①袁… ②刘… ③徐… ④张… ⑤刘 Ⅲ. ①饮食—文化—北京—汉、英 Ⅳ. ①TS971.202.1

中国国家版本馆CIP数据核字(2024)第004356号

馋了，风味儿北京（汉英对照）

袁小茶 文　刘天羽　徐翔宇 张伊聪 绘　刘海乐 译

出 版 人：方允仲
项目策划：徐和建
内容指导：谢　豫　周　齐
执行统筹：赵德璧
责任编辑：刘晓雪
英文编辑：王子木
英文改稿：（英）斯科特・亨斯迈
美术设计：王建东
内文排版：赵艳超
责任印制：焦　洋

出版发行：中国画报出版社
地　　址：中国北京市海淀区车公庄西路33号　邮编：100048
发 行 部：010-88417418　010-68414683（传真）
总编室兼传真：010-88417359　版权部：010-88417359

开　　本：16开（787mm×1092mm）
印　　张：14.75
字　　数：230千字
版　　次：2024年6月第1版　2024年6月第1次印刷
印　　刷：北京汇瑞嘉合文化发展有限公司
书　　号：ISBN 978-7-5146-2214-0
定　　价：108.00元

烤肉季

前言

北京既是一座拥有悠久历史的文化古城，又是一个活力四射的现代都市。

这里总会有一些令人惊奇的新发现。公园里，就着清晨第一缕阳光，老人将鸟笼挂在树枝上，悠然欣赏着婉转的鸟鸣；胡同口，青年麻利儿地装好刚出锅的油条，急匆匆地赶路；社区广场上，夕阳西下时，老年舞蹈队已经装备齐全，挥着大红扇子跳着广场舞……这些生活碎片无不洋溢着不同于其他城市的、颇具北京特色的活力。

“手绘北京”以北京的美食、休闲娱乐生活等内容为创作对象，通过手绘的艺术形式呈现北京的历史之韵、都市之美；以艺术家的视角，勾勒极具“东方特色”的北京风物；力求站在悠久历史与现代时空相交之处，来解读这座魅力城市。

艺术作品是最好的价值传播载体。坚持“美人之美、美美与共”，带您领略不一样的北京。

Foreword

Beijing is both an ancient capital with splendid culture and a modern metropolis with vibrant dynamics.

You can always find something new and interesting in this city. In a park, you can see seniors hanging their birdcages on the branches of the trees and enjoying the singing of their pet birds in the first rays of the morning sun; in an old alley, you can see young commuters rushing to work while agilely packing up the steamy deep-fried dough sticks they bought at a roadside food stall as breakfast; at a community square, well-dressed aged people dance energetically while swaying red fans in the setting sun… All of these radiate the unique vitality of Beijing, making it different from any other city.

The "Hand-Drawn Beijing" series focuses on depicting this city fusing tradition and modernity through hand-drawn illustrations of its cuisine, landmarks, entertainments, and others. From an artist's perspective, the book series presents a panoramic view of Beijing, a city sated with distinctive Eastern flavor. It seeks to interpret this charming city that integrates historical profundity and modern appeal.

Artworks are the best carriers for disseminating the common values of humanity. Following the principle of "upholding the beauty of each civilization and the diversity of civilizations in the world," the book series provides a unique view into Beijing.

目录
Contents

前言
Foreword

第一章 Chapter 1

犄角旮旯儿
Hidden Eateries

01 去尹三豆汁儿，拿塑料袋排队打一升“老北京的命根儿”
Buy Fermented Mung Bean Juice with a Plastic Bag at Yinsan Douzhir 002

02 去白记年糕，体验一回牛街排队最长的“驴打滚儿”
Queue Up to Taste the Yummiest *Ludagunr* at Bai's Bakery 004

03 去西海子西园烧饼夹肉，感受一次“味道好到不开外卖”的霸气
Experience “Chinese Hamburger” at the Xihaizi Xiyuan Restaurant 006

04 去“老回回”小吃，咬一口北京“天花板级”的糖花卷儿
Enjoy the Best Sweetened Steamed Rolls of Beijing at the Laohuihui Snacks 008

05 去宝瑞门钉肉饼店，空腹去啊，不然一个就饱了
Taste Doornail Meat Pie at the Baorui Restaurant 010

06 去万红路羊汤烧饼店（798 店），看艺术大咖们如何被一个土烧饼“征服”
Try Homemade Sesame Butter Flatbread at the Mutton Soup & Sesame Buns Restaurant near the 798 Art District 012

07 去门框胡同百年卤煮（新街口店），建议开车的朋友“腿儿着来吧”
Don't Drive to Menkuang Hutong Bainian Luzhu (Xinjiekou Branch) to Eat *Luzhu* 014

08 去沣元春饼馆，数数巴掌大的地方有多少张大腕儿合影
Count Celebrity Photos at the Fengyuanchun Spring Pancake Restaurant ... 016

09 宫门口馒头铺，咬一口“贝勒红豆卷”，当不了贝勒……
Taste Beile Red Bean Rolls at the Gongmenkou Steamed Bun Shop 018

10 去小肠陈，吃完卤煮，来个让人涕泗横流的“芥末墩儿”解腻
Try Pickled Mustard Chinese Cabbage after a Bowl of *Luzhu* at Xiaochang Chen's Restaurant 020

第二章 Chapter 2

八大楼、八大居和四大顺
Famous Time-Honored Restaurants

01 去东兴楼，尝一回“八大楼”之首的美食

Enjoy Delicacies at Dongxing Lou, Top of the “Eight Lou” Restaurants . 024

02 去泰丰楼，体验一回“唱戏的腔，厨师的汤”

Taste the Chef's Soup Blending Diverse Flavors at Taifeng Lou 026

03 去致美楼，感受一次“御厨遗风”

Experience the Delicacies of the “Imperial Kitchen” at Zhimei Lou........ 028

04 去萃华楼，掰一口“特色烤花卷”

Try Roasted Twisty Rolls at Cuihua Lou 030

05 去砂锅居，干瞪眼看对面小哥“包圆儿”砂锅白肉

Enjoy Pork and Vegetable Stewed in Chinese Clay Pot at Shaguo Ju 032

06 去同和居，享受一回当年鲁迅、齐白石都来的老字号

Enjoy Dinner at Tonghe Ju, a Restaurant Frequented by

Lu Xun and Qi Baishi.. 034

07 去柳泉居，体验一回舌尖上的“北京市非遗”京菜技艺

Tickle Your Taste Buds with Liuquan Ju's Intangible Cultural

Heritage Delicacies 036

08 去东来顺，小料一定得咬牙要麻酱底儿

Choose Sesame Paste Dipping Sauce for Hotpot at Donglai Shun 038

09 去南来顺，感受一次地道南城味道

Experience the Authentic Delicacies of Southern Beijing at

Nanlai Shun... 040

10 去西来顺，点一份“马连良鸭子”

Taste Ma Lianliang Duck at Xilai Shun 042

11 去又一顺，皇亲国戚曾是座上宾

Dine at Youyi Shun like the Royals 044

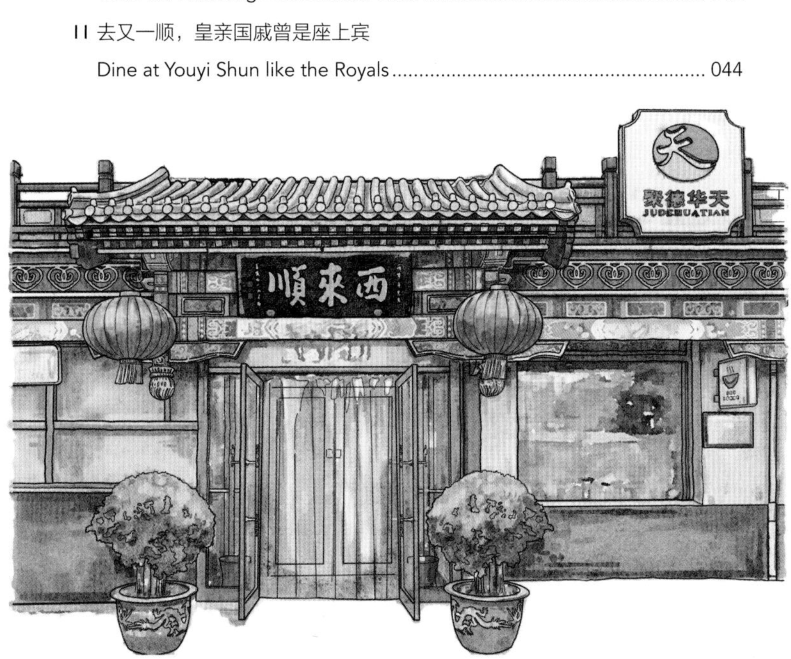

第三章 Chapter 3

老国营 老味道
State-Run Restaurants and Groceries

01 去赵府街副食店，拎着瓶子来打散装“二八酱”
Buy Bulk Mixed Paste at the Zhaofujie Grocery 048

02 去小楼饭店，一定得来一份烧鲇鱼
Try Catfish Braised in Soy Sauce at the Xiaolou Restaurant 050

03 去大顺斋，寻觅老国营的糖火烧
Buy Sugared Sesame Buns at the State-Run Dashun Zhai 052

04 去烤肉季（什刹海总店），体验一回“武吃”
Experience "Martial Dining" at the Shichahai Flagship Store of Kaorou Ji .. 054

05 去锦芳小吃，尝一回老北京的什锦元宵
Taste Assorted *Yuanxiao* of Old Beijing at the Jinfang Snack Store 057

06 去新成炸酱面馆，数数一碗炸酱面里有多少块儿肉丁儿
Count Diced Pork Pieces in a Bowl of Noodles with Soybean Paste at the Xincheng Noodle Restaurant 058

07 去天兴居，体验一把“喝炒肝儿”的老规矩
Eat *Chaogan* the Traditional Way at Tianxing Ju 060

08 去丰春楼饭庄，尝一口老国营的焦熘丸子
Try Deep-Fried Meatballs at the State-Run Fengchunlou Restaurant 062

09 去北海公园仿膳饭庄，尝一回真正“宫里的菜”
Taste Authentic Royal Cuisine at the Fangshan Restaurant in Beihai Park 064

10 去百年义利，找一口儿时的维生素面包
Recall Your Childhood by Tasting Vitamin Bread at the Century-Old Yili Bakery 066

11 去新新包子铺，尝一回“两元一个，只卖一种馅儿”的霸气包子
Buy Steamed Buns Stuffed with Pork and Scallion at Xinxin 068

12 去杏园餐厅，吃一次“不带滤镜”的老北京糖醋里脊
Enjoy Authentic Beijing-Style Sautéed Sweet and Sour Pork Tenderloin at the Xingyuan Restaurant 070

第四章 Chapter 4

中外闻名北京味儿
Beijing Flavors with Global Fame

01 去利群烤鸭店，尝一尝胡同儿四合院里的北京果木烤鸭
Taste Beijing Roast Duck at the Liqun Restaurant Hidden in an Alley 075

02 去富华斋饽饽铺，尝一尝《红楼梦》里的宫廷饽饽
Taste Royal Steamed Buns Depicted in *Dream of the Red Chamber* at Fuhuazhai 076

03 去大董，体验一回“黑珍珠三钻”的北京烤鸭
Eat Roast Duck at Dadong, a “Once in a Lifetime” Restaurant 078

04 去“提督”，品尝烤鸭配西式鱼子酱
Experience Roast Duck with Caviar at TIDU 080

05 去方砖厂 69 号，尝一尝米其林一星的“北京炸酱面”
Enjoy Beijing-Style Noodles with Soybean Paste at MICHELIN Star-Rated Restaurant Fangzhuanchang No. 69 082

06 去四季民福，尝一回口碑相传的“焖炉烤鸭”
Taste the Duck Roasted in a Closed Oven at the Peking Chamber 084

07 去宝源饺子屋，宫保鸡丁也能做饺子馅儿?
Experience Dumplings Stuffed with Kung Pao Chicken at Baoyuan 086

08 去金生隆，天啊！北京爆肚儿还带中西结合的?
Enjoy Boiled Tripe Fusing Chinese and Western Styles at Jinshenglong 088

09 去牛街满恒记，体验一回老北京铜锅涮肉的“阵势”
Experience Beijing-Style Copper Hotpot at Manhengji on Niujie Street 090

10 去前门六必居，约半斤老北京酱菜“麻仁金丝”
Buy Pickled Kohlrabi with Sesame Seeds at Liubiju in Qianmen 092

第五章 Chapter 5

来都来了
Home-Style Eateries

01 去后海的鸦儿李记，一口烧饼咬下去，成仙了

Enjoy Baked Sesame Buns at Ya'er Liji on the Shore of Houhai Lake...... 097

02 去南锣鼓巷的“菊儿人家”，体验一回“北京最牛十碗饭之首”

Experience Beijing's Best Rice with Stewed Pork at Ju'er Renjia in Nanluoguxiang 099

03 去香饵胡同的胖妹面庄，尝一回米其林“必比登”推荐的辣面

Try the Spicy Noodles Rated in the MICHELIN Guide from the Pangmei Noodles Restaurant in Xiang'er Hutong 100

04 去吃一回老北京的“局气”

Experience Creative Beijing Cuisine at Juqi 102

05 去报房胡同饺子馆，体验一回“躺平饺子”

Taste Homemade Chinese Dumplings in Baofang Hutong 104

06 去雍和宫大街的新和小馆，恭喜您排上号了

Compete to Get a Seat at the Xinhe Restaurant on Yonghegong Street 106

07 去东四民芳餐厅，尝一回老北京发小儿“重口味零食”

Awaken Childhood Memories at the Dongsi Minfang Restaurant........... 108

08 去天福号，称半斤乾隆三年传下来的“老北京酱肘子”

Buy Braised Pork Hock with Soy Sauce at the Century-Old Tianfuhao.... 110

09 去紫光园，尝一回镇店的百年京菜“炒疙瘩”

Taste Stir-Fried Starch Knots at the Time-Honored Ziguangyuan Restaurant 112

10 去平安小馆儿，尝一回北京老百姓口味的地道锅贴儿

Taste Authentic Beijing-Style Pan-Fried Dumplings at Pingan Xiaoguan 114

11 去大华煎饼，尝一回老北京胡同里的“薄脆天花板”
Taste the Best Pancakes at Dahua, Hidden in a Beijing Lane 116

12 去新桥炸鸡店，点一份儿时味道的“老式炸鸡”
Experience Traditional Fried Chicken from Childhood at the Xinqiao Fried Chicken Restaurant 118

13 去大秤钩胡同，尝一回地道的“老北京馅饼粥”
Taste Authentic Beijing-Style Pie and Porridge in Dachenggou Hutong 120

第六章 Chapter 6

瓷，咱吃的不是菜，是“意境”
More than Just Food

01 去和芳苑，在四进四合院的榜眼府第，喝一回下午茶
Enjoy a Traditional yet Fancy Afternoon Tea in the Courtyard of Hefangyuan... 124

02 去曲廊院，感受一回四合院里的“中西交融艺术菜”
Experience Artistic Dishes Fusing Chinese and Western Styles at Qulangyuan 126

03 去望春食阁，景山下尝一回“宫墙四重奏”下午茶
Experience the “Royal Palace Quartet” at Wangchun Shige at the Foot of Prospect Hill 128

04 路过奥体中心，尝一次梦回《东京梦华录》里的“宋宴”
Experience Song Dynasty Lifestyles Described in *The Dreamlike Glory of the East Capital* at the Song Banquet 130

05 去“宫宴”，穿着汉服体验一回《礼记》里的燕享食礼
Experience the Dining Ritual Recorded in *The Book of Rites* at the Palace Banquet 132

06 去白家大院，尝一回旧王府的“格格宫廷菜”
Taste Royal Cuisine at Bai's Compound, a Former Prince's Mansion....... 134

07 去御仙都皇家菜博物馆，尝尝“皇家菜”的非遗传承
Enjoy Cultural Heritage at Yuxiandu Imperial Cuisine Museum............... 137

08 去副中心的湿地公园，走一趟“能吃鱼的博物馆”
Eat Fish in a Museum-like Restaurant Hidden in a Wetland Park 138

09 去北平楼，重回晚清民国风的“北平民间菜”
Experience Folk Cuisine from the Late Qing Dynasty and Early Republic of China Period at the Beipinglou Restaurant 140

10 去和平果局老北京风物馆，穿越 40 年前的“老北京胡同儿”吃食
Revive Old-Time Beijing Lifestyles at Heping Guoju............................... 142

第七章 Chapter 7

肉食动物
Paradises for Meat Gourmets

01 去烤肉宛，“文吃”师傅的手艺同样“豪气”
Experience "Literary Dining" at Kaorou Wan 146

02 去烤肉刘，找找汪曾祺散文里的“主人公”
Rediscover the Protagonists in Wang Zengqi's Prose at Kaorou Liu........ 148

03 去柴氏风味斋，吃一碗米其林“必比登”榜上的“小碗肉”
Eat Small Beef Bowl Rated in the MICHELIN Guide at Chai's Fengweizhai 150

04 去锡拉胡同，尝一回满满肉馅的“Beijing Pie”（“北京肉饼”）
Taste Meaty Beijing Pie in Xila Hutong... 152

05 去“鼓楼吃面”，来一盘摇滚味儿的“肉啃肉”
Experience "Rock and Roll" and Meaty Foods at Gulou Chimian........... 155

06 去合兴楼，尝一回“御膳肘子王”
Taste King's Pork Hock at Hexing Lou 156

07 去群生大院，尝一回 1910 年的老字号“东坝驴肉”
Enjoy Dongba Donkey Meat in the Time-Honored Restaurant
the Qunsheng Compound 158

第八章 Chapter 8

爷今儿吃素
Going Vegetarian for One Day

01 去京兆尹，吃一顿纯素的“米其林三星”
Enjoy Vegetarian Foods at the King's Joy, a Three-Starred Restaurant
in the MICHELIN Guide 162

02 去山河万朵，尝一回“素食天花板”的艺术家仙气
Taste Artistic, Immortal Vegetarian Foods at the Vege Wonder............. 164

03 去叶叶菩提，尝尝二十四节气里的“百味养生”
Experience Herbal Foods Based on the 24 Solar Terms at the Yeye Bodhi.... 166

04 去静莲斋，体验一回平民消费的“素菜荤吃”
Enjoy Affordable Vegetarian Foods at Jinglianzhai 168

05 去素虎，感受一回“菜虎子”心心念念的素食自助
Enjoy Vegetarian Buffet at the Vege Tiger Restaurant............................ 170

06 去花开素食，尝尝人均消费不到二百块的米其林餐厅！
Dine at the Blossom Vegetarian, a MICHELIN Restaurant with a
Per Capita Spending of Less than 200 Yuan 172

07 去福慧慈缘素食餐厅，胃安静了，心就放松了
Satisfy Your Stomach with Mindful Eating at the Fuhuiciyuan Vegetarian
Restaurant 174

08 去东草园胡同，感受一次“藏茶素火锅”

Experience the Tibetan Tea Vegan Hotpot in Dongcaoyuan Hutong 176

09 路过北大，体验一回“草木禅房”的养生素食

Experience Healthy Vegetarian Diets at Caomu Chanfang near Peking University 178

第九章 Chapter 9

好这口儿
Favor This Flavor

01 去三里屯“姆们”，咬一口“人生苦短，姆们很甜”

Sweeten Your Life at the M Sweetie Cake in Sanlitun............................. 182

02 去元古云境，尝一口“颜值甜点”茉莉清茶酪

Taste Jasmine Tea Yogurt at Yuangu Yunjing 184

03 去雁舍，吃一回 8 岁到 80 岁都能吃的“辣菜”

Eat Spicy Foods Suited for Diners Aged between Eight and Eighty at Yanshe 186

04 去麻六记，记得早晨 10 点排队“堵门”拿号

Queue Up to Wait for a Seat at Maluji as Soon as It Opens at 10 A.M...... 188

05 去贵州大厦，感受一回“驻京办”的地道酸汤鱼

Taste Authentic Guizhou-Style Fish in Sour Soup in the Guizhou Hotel....... 190

06 去岐黄小馆，吃一回有发明专利的“养生药膳”

Experience Patented “Medicinal Diet” at Qihuang Xiaoguan 192

07 去郭林家常菜，见一回“好久不见”的老朋友

Reunite with Old Friends at the Guolin Restaurant 194

08 去后永康胡同，到“做书咖啡”点杯美式撸个猫

Sip a Cup of Americano while Playing with Cats at the Zuoshu Café 196

第十章 Chapter 10

我爱驻京办

Delicacies Hidden in Beijing Offices of Local Governments

01 去南宁大厦嗦一碗老友粉，据说连建筑都腌入味了
Taste Old Friends Rice Noodles at the Guangxi-Style Restaurant in the Nanning Hotel 200

02 去宜宾招待所点一回地道川菜，感觉前半生吃过的“鱼香肉丝”都白吃了
Enjoy Authentic Sichuan Cuisine at the Yibin Guesthouse in Beijing 202

03 去重庆饭店，辣子鸡的辣椒都可以打包的水准
Pack the Leftover Chili Peppers after Enjoying Spicy Chicken in the Chongqing Hotel 204

04 去湖北大厦，来一份“神仙水”级别的莲藕排骨汤
Taste the Yummy Pork Chop Soup with Lotus Roots in the Hubei Hotel 206

05 去安徽大厦，点一份惊艳的臭鳜鱼
Taste the Amazing Smelly Mandarin Fish in the Anhui Tower 208

06 去湖南大厦，数一数剁椒鱼头能下几碗米饭
Enjoy Steamed Fish Head with Diced Hot Red Peppers in the Hunan Hotel........ 210

07 去云腾食府，从此爱上云南的米线和各种蘑菇
Fall in Love with Yunnan Rice Noodles and Mushrooms after a Visit to Yunteng Shifu........ 212

08 去赣人之家，报恩的瓦罐汤和“报仇”的辣
Taste Jiangxi-Style Clay Pot Soup at Ganren Zhijia 214

09 去河南大厦，干一碗魂牵梦绕的羊肉烩面
Taste a Long-Waited Bowl of Stewed Noodles with Mutton in the Henan Hotel........ 216

10 去浙江大厦，江南不仅有“西湖牛肉羹”
Taste West Lake Beef Soup and More in the Zhejiang Hotel 218

如果您来北京有一阵子了，还徘徊在上顿“全聚德”、下顿“大董”，这并不能说明您懂“吃”。只能证明三点：第一，您不差钱；第二，您可能不太懂北京；第三，您不太懂生活。

一个城市真正的人间烟火味，是在那些犄角旮旯儿、火到让你怀疑人生的“小门脸儿”里。

以下推荐 10 个“小门脸儿”打卡地。

Living in Beijing for an extended period while still frequenting famous restaurants such as Quanjude and Dadong doesn't make one a gourmet, but indicates three things: first, the person is wealthy; second, he or she might know hardly anything about Beijing; third, he or she leads an uptight lifestyle.

A city's signature delicacies are usually tucked away in time-honored eateries that look shabby but remain extremely popular among locals. Though small and crowded, such places in modest corners of the city are definitely worth visiting.

The following are ten such hidden eateries we recommend.

第一章
Chapter 1

犄角旮旯儿

Hidden Eateries

01

东城区东晓市街 176 号 · 176 Dongxiaoshi Street, Dongcheng District

去尹三豆汁儿，拿塑料袋排队打一升“老北京的命根儿”

Buy Fermented Mung Bean Juice with a Plastic Bag at Yinsan Douzhir

“北京的豆汁儿，旗人的命根儿。”您要是路过天坛附近的东晓市街，看到零下十几度的寒风里一群大爷大妈抱着胳膊根儿排大队买早点，那都不用再往前走看招牌了，肯定是尹三豆汁店——生生把一碗豆汁儿做成上榜米其林“必比登推荐”的小吃店。他家的豆汁儿是重发酵——要的就是这又稠又酸的味儿。

在尹三豆汁店，如何判断一个真正的“老北京”？您看那一碗豆汁儿有95%都给手机“喝了”的、龇牙咧嘴的，肯定是网红观光客，真正附近的老北京都是自带塑料袋买生豆汁儿、生麻豆腐——生的价格实惠，可以论斤约（yāo，北京话，论斤称量的意思）。北京大爷嘴里还得扯着嗓门嘱咐一句：“把那豆汁儿给我搅和搅和，别净给我盛那稀汤寡水儿的。再给我来三袋儿！”

A saying goes, "Beijing douzhir is the lifeline of local Manchu residents." *Douzhir*, a kind of fermented mung bean juice, is popular among Beijingers. If you pass Dongxiaoshi Street near the Temple of Heaven in winter, you'll see a group of elderly people lining up in front of a store despite chilly winds. It is Yinsan Douzhir, a local eatery rated in the MICHELIN Guide. The *douzhir* it sells is heavily fermented and tastes thick and sour.

How does one distinguish between tourists and regulars of the restaurant? Someone wincing when sipping a cup of *douzhir* while happily taking a selfie is certainly a tourist. Local residents usually take uncooked *douzhir* or *madoufu* (a traditional Beijing snack made of cooking starch residue) away in a plastic bag. The uncooked version is much cheaper, and people can cook it at home whenever they want. "Stir the *douzhir* for a while to make it thick or it will be watery and tasteless!" shouted an old man to the cook. "Give me three bags!"

尹三豆汁店
豆汁

02

西城区牛街街道牛街 5 号牛街清真超市 1 楼（近输入胡同）
F/1, Niujie Halal Supermarket (near Shuru Lane), 5 Niujie Street, Xicheng District

去白记年糕，体验一回牛街排队最长的“驴打滚儿”

Queue Up to Taste the Yummiest *Ludagunr* at Bai's Bakery

在牛街，有两条“神奇”的队——不论刮风下雨，永远“乌泱乌泱”地人挤人，跟不要钱似的。一条是排队等号吃聚宝源火锅的，另一条就是排队等白记年糕买小吃的：豌豆黄儿、艾窝窝、盆儿糕、糖火烧，还有必买的招牌“驴打滚儿”——北京人童年记忆里最甜最糯的点心之一。之所以得了一个这么不登大雅之堂的名儿，据传是因为“驴打滚儿”最后一道工序，是要在黏豆面糕的表面裹一层黄豆面儿，像北京郊外的野驴撅蹄子扬的黄土。

如果您是头一回来，友情提醒您千万别“贪多”——因为这些甜食都是黏米面儿做的，小时候家大人（北京话：家长的意思）都不许多吃，怕不消化积食。而且黏米面儿最好不隔夜，吃的就是一个趁热的软糯劲儿，一放冰箱就干了，味道差一半，买多了肯定糟践了。当然了，我说这些也肯定没用——来这儿排队的都是每样点心半斤起步，不然对不住自己排了俩小时队啊！

On Niujie Street, customers are always queuing up outside two stores: one is the time-honored hotpot restaurant Jubaoyuan, and the other is Bai's Bakery, which offers a plethora of Beijing snacks including *wandouhuang* (pea flour cake), *aiwowo* (steamed rice cake with sweet stuffing), *pen'ergao* (pea cake steamed in pottery basin), *tanghuoshao* (baked wheaten cake with sugar), and of course, its signature offering—*ludagunr* (glutinous rice rolls with sweet bean flour). *Ludagunr*, or Rolling Donkey, is perhaps one of the sweetest snacks popular among native children in Beijing. Why did the snack get such a vulgar name? The last step to prepare the snack is spraying bean flour on the surface, which looks like yellow-brown dirt stirred up by a donkey rolling in the wild.

When you first try *ludagunr*, don't eat too much because it is made of glutinous rice, which is not easy to digest. Also, you'd better eat it as soon as you can. It only tastes soft and sweet soon after cooking. Its flavor will deteriorate after one night of storage in a refrigerator. So, don't buy too much one time or they will go to waste. Nevertheless, most customers buy at least 250 grams of each snack after waiting for two hours in line.

記年糕
白記年糕

03

通州区云景东路 339 号 · 339 Yunjing East Road, Tongzhou District

去西海子西园烧饼夹肉，感受一次“味道好到不开外卖”的霸气
Experience “Chinese Hamburger” at the Xihaizi Xiyuan Restaurant

来，先跟着我念一个发音：西海子（zi，轻声）。老北京方言受到蒙元文化影响，管“大湖”叫“海子（zi，轻声）”。您要是到了副中心，问“西海子（zǐ）怎么走”，老通州可能一愣，在热心给你指路的同时，不忘上一堂生动的“方言课”。西海子的烧饼夹肉，也是一家有几十年历史的低调“老国营”——就卖烧饼夹肉。味道好到什么程度呢？人家根本不开“外卖”，更不会像“网红店”一样整一堆花里胡哨的餐具求好评。您想吃这口儿，对不起，劳驾自己来排队，或者加钱雇网络“跑腿儿”。2020年之前，人家连微信支付都不用——只收现金。既使这样，每天依然是乌泱乌泱的人，来晚了肘子肉就没了。这也公平——无论您祖上是三公九卿，还是平头百姓，在等待一锅热烧饼出锅的刹那，平等了。

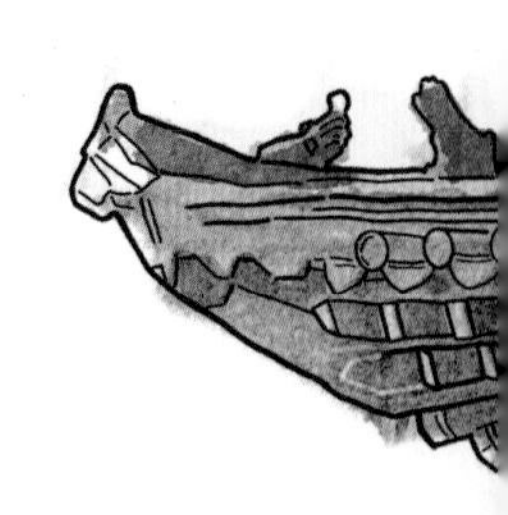

Beijing dialect was influenced by the Mongolian culture during the Yuan Dynasty (1206-1368), and locals call lakes *hazi* (meaning “sea”). If you ask a senior native of Tongzhou District, Beijing's sub-center, for the way to Xihaizi, he or she may gladly offer help by teaching you about Beijing dialect. Xihaizi Xiyuan is a decades-old state-run restaurant, and the only product it offers is *shaobingjiarou* (meat-sandwiched pancake), nicknamed “Chinese hamburger.” The snack is so popular that the restaurant doesn't offer delivery service. To grab a bite, one must come to the restaurant in person or hire someone to queue for him. Before the outbreak of COVID-19, the restaurant only took cash, and customers couldn't pay with WeChat. Even so, it was packed with customers every day, and latecomers would have to return home empty-handed. It serves everyone from anywhere, regardless of social status. All are equal when they line up to wait for a steamy *shaobingjiarou*.

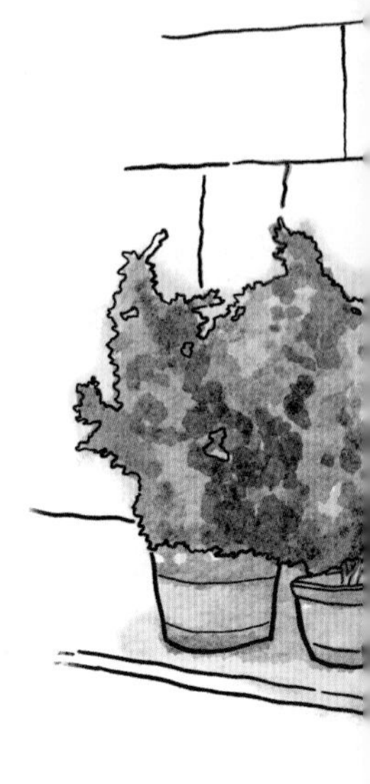

烧餅夹肉

04

西城区百万庄北街9号1层 · F/1, 9 Baiwanzhuang North Street, Xicheng District

去“老回回”小吃，咬一口北京“天花板级”的糖花卷儿

Enjoy the Best Sweetened Steamed Rolls of Beijing at the Laohuihui Snacks

“北京小吃的半壁江山，都是清真小吃。原本是附近居民口碑相传的小吃店，因为“天花板级”的麻酱红糖花卷儿，征服了四海八方年轻人的味蕾，一举成为“网红小吃打卡地”。其实老北京的早点没那么多甜的，咸豆腐脑才是正确的打开方式。豆腐要老嫩适中，卤要用牛肉汤打底，黄花、木耳、香菇一应齐备。早晨来碗“老回回”的豆腐脑，配点儿什么呢？要不来个糖油饼儿？又是甜的。——碳水让人快乐，糖油饼儿双倍碳水，双倍快乐。

Actually, nearly half of Beijing snacks are Halal foods. At first, the Laohuihui Snacks eatery was only popular among local residents. Eventually, sweetened steamed rolls with sesame sauce, considered the best in Beijing by many, were attracting young diners from all over China after the eatery became an “Internet celebrity.” Beijing's traditional breakfast snacks don't have much sugar, and salty jellied bean curd is the most popular breakfast snack in the city. The bean curd must be mildly tender, and accompanied by beef soup and seasoning like dried daylily, sliced wood ear mushroomsr, and shiitake mushrooms. In addition to a bowl of jellied bean curd, what else can you order at the Laohuihui Snacks? Maybe a deep-fried dough cake. It's also sweet. Carbs lead to happiness, and a sweet deep-fried dough cake can double it.

老回回小吃
传统碗菜 棒骨火锅
老回回
来的百年老号
熟食酱货

05

东城区朝阳门北小街61-1号 · 61-1 Chaoyangmen Beixiaojie, Dongcheng District

去宝瑞门钉肉饼店，空腹去啊，不然一个就饱了

Taste Doornail Meat Pie at the Baorui Restaurant

“门钉肉饼”为什么叫“门钉肉饼”呢？——因为形状长得像故宫午门的门钉，圆乎又瓷实，比手小的女生拳头还大。去宝瑞吃门钉肉饼，吃的就是他家皮儿薄馅儿大，最外层的面皮儿煎得微焦发黄，里边的肉馅儿汤汁浓郁，不腥不柴，又解馋又管饱。吃门钉肉饼有讲究，一定得趁热、小口咬——不然汤汁烫嘴，大口咬下去“爆汁儿”，容易滋（北京话，溅的意思）一身油。得，羽绒服送干洗店吧，肯定比7块钱一个的肉饼贵。

宝瑞门钉肉饼虽好，也不建议您一次点太多。馅儿太“瓷实”了，尤其是保持身材“小鸟儿胃”的女生，基本上一个下去——饱了。另外，门钉肉饼的面皮儿您可千万别糟蹋，肉馅儿的汤汁精华都沁在这薄面皮儿上了，再蘸一口醋解腻，嘿，还能再吃一个。

Doornail meat pie was so named because it looks like the doornail decorating the Meridian Gate of the Forbidden City in Beijing. Round and solid, it is bigger than a girl's fist. Doornail meat pie from the Baorui Restaurant features thin crust and plenty of filling. The outside crust is yellowish and crispy, while the stuffing is juicy and chewy. A bite can generate a sense of satiety. Advice on eating doornail meat pie: attempt small bites to avoid the hot juicy filling bursting out and soiling your shirt. The price of dry-cleaning is definitely more than a doornail meat pie, at 7 yuan.

Although doornail meat pie is appetizing, don't eat too much at once. Its filling is so meaty and thick that any dieter will call it quits after one. Also, don't waste the thin crust, which is also lip-smacking. Dipping the pie in vinegar can cleanse the palate, after which you may want another.

宝瑞门钉肉饼店

06

朝阳区万红路5号蓝涛中心 · Lantao Center, 5 Wanhong Road, Chaoyang District

去万红路羊汤烧饼店（798店），看艺术大咖们如何被一个土烧饼“征服”

Try Homemade Sesame Butter Flatbread at the Mutton Soup & Sesame Buns Restaurant near the 798 Art District

如果您对798艺术街区的刻板印象，还停留在一身潮牌儿的“当代艺术家范儿”，那就落伍了。比798艺术展的队更火的，是门口卖土烧饼和羊汤门前的队……赶上中午饭点儿，还以为这是啥“人间至味”呢，乌泱泱排队的人群里不乏背着“爱马仕”的小姑娘和留着胡子的艺术大叔——就着北京春天“土大”的风，一边啃着土烧饼，一边跟您接着聊本雅明“机械复制时代的艺术”。

为什么这么多人排队呢？因为太火了。店家只好在高峰时期限购，烧饼每个人最多买两个。他家的烧饼用料十足，价格又公道，趁着热一口咬下去，刚出锅的麻酱甜咸味儿混着烧饼的层次感在口腔里翻涌，让一颗前一秒还在算“碳水化合物卡路里”的减肥心瞬间被征服，算了，我现在不想听博伊斯《如何向一只死兔子解释绘画》，只想搞一个《如何向博伊斯解释中国“烧饼”》。

You're behind the times if you still consider the 798 Art District just a gathering place for contemporary artists and their works. These days, a down-to-earth restaurant specializing in homemade sesame butter flatbread and mutton soup draws more people than any art exhibition. At lunchtime, you'll find trendy ladies holding Hermès bags and middle-aged artists with beards in the long line outside the restaurant. Especially in Beijing's windy spring, some are stunned by a scene of two contemporary artists discussing Walter Benjamin's essay "The Work of Art in the Age of Mechanical Reproduction" while chewing sesame butter flatbread.

Why is the line always long? The restaurant's sesame butter flatbread is so popular that it has to limit each customer to only two per purchase.

The sesame butter flatbread it offers has more sesame butter and features more reasonable prices. Just a bite of the sweet, salty sesame butter tickles the taste buds and conquers the heart. Your worries about calorie intake will melt away. Instead of listening to German artist Joseph Beuys's *How to Explain Pictures to a Dead Hare*, you may want to explain Chinese sesame butter flatbread to him.

07

西城区赵登禹路甲 2 号（新街口地铁 B 口正对面）· A2 Zhaodengyu Road (opposite Exit B of Xinjiekou Subway Station), Xicheng District)

去门框胡同百年卤煮（新街口店），建议开车的朋友“腿儿着来吧”

Don't Drive to Menkuang Hutong Bainian Luzhu (Xinjiekou Branch) to Eat *Luzhu*

在四九城生活几年，您可能会添一个“坏毛病”——爱上吃“下水（shui 轻声）”了。老北京管动物内脏叫“下水”。卤煮里最重要的两味材料，就是猪肠和猪肺。吃卤煮这件事，有点儿像蔬菜界的“香菜”、小吃界的“臭豆腐”、饮料界的“豆汁儿”——总之就是能吃惯的人嗜此如命、无此不欢，吃不惯的人骂骂咧咧、避之大吉。吃卤煮的习惯见仁见智，有人会跟老板说“不要肺头”，也有“老饕”觉得“肺头”才是一碗卤煮的“完整尊严”。

不过要提醒您的是，新街口店虽然“味儿正”，但真的不好停车，带朋友来的话，一定慎重选择出行方式。这里完全符合您对一个网红小店的期待：店里找不到座儿，店外没地儿停车。建议开车的朋友找个远点儿的停车场“弃车前行”——“腿儿着来吧”。

After living in Beijing's old district for several years, many build a new habit of eating *xiashui* ("animal offal" in Beijing dialect). The two most essential ingredients of *luzhu*, a traditional Beijing snack, are pork lungs and intestines. Just like coriander, fried stinky tofu and *douzhir*, fans of *luzhu* tend to be obsessed with it while its detractors won't touch it. Some order it without pork lungs, but traditionalists would consider that non-authentic.

Despite its extraordinary *luzhu*, the Xinjiekou Branch of Menkuang Hutong Bainian Luzhu doesn't offer on-site parking. Driving there is never a good option. Hidden in the crowded old district, the restaurant meets all expectations for an Internet-famous eatery—small, old but popular. In addition to the lack of parking, there is a wait during peak hours. If you have to drive, park several blocks away and then walk there.

胡同百年卤煮

08

东城区幸福巷幸福北里 20 号楼 · Building 20, Xingfu Beili, Xingfu Lane, Dongcheng District

去沣元春饼馆，数数巴掌大的地方有多少张大腕儿合影

Count Celebrity Photos at the Fengyuanchun Spring Pancake Restaurant

一家藏在居民楼里的春饼馆子，居然火得很“邪性”。如果您按照“商学院高端授课”的思路，这应该属于“基本歇菜”的反面教材——选址不临街，得靠导航才能找来；菜也不是什么稀罕菜系，就是最平常不过、老北京家家户户都能做的春饼。嘿，但人家老板就是扎扎实实把春饼做火了。一个巴掌大的馆子，墙上一大堆“明星合照”，就为了这一口老北京春饼。

来这儿吃老北京春饼，必点的一道菜就是“炒合菜”。其实合菜可荤可素，过去老北京家家户户的配料表都不尽相同，但大致都不出老北京春天的几样常见时令菜：豆芽儿、韭菜、鸡蛋、木耳、粉丝。过去“反季节蔬菜”很少，北京的春天“青黄不接”，这一筷子炒合菜卷上春饼，就算是北京人“咬春”了。要是嫌干，可以来一碗棒渣粥“溜缝儿”。

Despite being tucked away in a residential building the Fengyuanchun Spring Pancake Restaurant is surprisingly popular. The restaurant is the polar opposite of successful cases offered in textbooks for business colleges: it isn't close to any main street, and a first-time visitor must depend on a navigation app to get to the restaurant; its signature offer, spring pancakes, is common on local dining tables. Yet, the restaurant has been a great success, evidenced by many photos of the celebrities frequenting it on the walls.

One of the restaurant's signature dishes is stir-fried assorted vegetables. Despite its name, the dish may also contain meat. In fact, the recipe for this dish varies in different families in Beijing. The most common ingredients include mung bean sprouts, Chinese leeks, eggs, wood ear mushrooms, and glass noodles. In the past, only a few varieties of vegetables were available to Beijing residents in early spring. Locals would wrap fried assorted vegetables with a spring pancake as a food for the Beginning of Spring, one of the 24 solar terms in the Chinese lunar calendar, and the tradition is called "biting the spring." The dish pairs well with a bowl of *bangzha* (minced corn) congee.

沣元春饼馆
沣元春饼馆

09

西城区香厂路 1 号 · 1 Xiangchang Road, Xicheng District

宫门口馒头铺，咬一口“贝勒红豆卷”，当不了贝勒……

Taste Beile Red Bean Rolls at the Gongmenkou Steamed Bun Shop

“宫门口”是一家卖馒头起家的主食小店，火到排队能排出二里地去。它家最火的全是减肥和控糖界的天敌“碳水炸弹”——贝勒红豆卷、八宝黏豆包、抹茶馒头……翻译过来吧，全是“主食”。

头一回看到宫门口“红糖麻酱花卷”的价目表一愣，以为是什么新品种，其实就是老北京的糖花卷儿。二十世纪九十年代初的时候，老北京家家户户别说“空气炸锅”，连有“烤箱”的人家都是屈指可数，大部分的妈妈们能给孩子在饭桌上变点花样儿的“甜点心”，就是一口蒸锅蒸出来的“糖花卷儿”——一定要多放红糖、多放麻酱，出蒸锅恨不得能从花卷周边溢出来，溜边儿溜沿儿的糖酱配着花卷的层层叠叠，冬天一口咬下去，眼镜片起了白色的雾气，朦朦胧胧像是回到了小时候，那时候的妈妈浓密乌黑的头发，能梳两个握不住的麻花辫，我还是乳牙未退的“豁牙子”，碗里多一个糖花卷儿就是世界上最快乐的事。不行，再写下去眼镜片儿要花了，“老板，两个糖花卷儿，打包”。

Gongmenkou is a small shop that first specialized in steamed buns. Waiting customers often form a line as long as a kilometer, testifying to the astonishing popularity of the shop. In fact, all the commodities it offers are high-carbohydrate foods such as Beile red bean rolls, Babao bean buns, and steamed matcha buns, which might be “nightmares” for those who aspire to lose weight.

The first time I saw “Twisted Rolls with Brown Sugar and Sesame Butter” on its menu, I mistakenly thought it was a new offer. Actually, the product is what is more commonly known as “steamed sugar rolls.” Back in the 1990s, few families in Beijing had an oven, let alone air fryers. In this context, steamed sugar rolls were one of the few sweet pastries that a mother could prepare for her kids. The secret to cooking sugared rolls is to add as much brown sugar and sesame butter as possible. When the steamed rolls are first ready, the sugar and butter seem to overflow. In winter, when taking a bite of such just cooked rolls, the steam blurred my glasses, giving me the illusion that I was eating steamed sugar rolls cooked by my mother during my childhood, who was wearing two large braids back then. At that time, nothing else could make me feel happier. My glasses would become more blurry if I continued eating. “Please pack up two more sugared rolls, and I'll take them to go,” I told the proprietor.

宫门口
宫门口馒头铺
—创始人店—

西城区南横东街 194 号 · 194 Nanheng East Street, Xicheng District

去小肠陈，吃完卤煮，来个让人涕泗横流的“芥末墩儿”解腻

Try Pickled Mustard Chinese Cabbage after a Bowl of *Luzhu* at Xiaochang Chen's Restaurant

小肠陈本是1886年创办的老字号，排队的食客都是慕名吃卤煮的。吃完卤煮火烧别一抹嘴儿走了，讲究的“老北京”都得来份儿芥末墩儿解腻。

芥末墩儿也是北京小吃中的一朵“奇葩”，杀伤力不亚于豆汁儿和臭豆腐。很多外省或外国朋友来北京吃饭，芥末墩儿基本上“这辈子尝过一筷子就够了”，辣得鼻涕眼泪直流，但“老北京”却对芥末墩儿乐此不疲。到底什么是芥末墩儿呢？其实材料便宜得要命，就是大白菜、黄芥末和一点儿糖、醋，凉凉的白菜帮加上芥末蹿鼻的味道，吃后真如醍醐灌顶一般，彻头彻尾地通气清凉。这道菜在过年或者肉菜多的时候，是老北京餐桌上最常见的配菜——就靠它醒脾解腻。小肠陈的卤煮讲究汤底醇厚，肠子里要稍微带点油才有味儿，不能太寡淡。这时候您稍微配一口芥末墩儿，会觉得从心里到胃里都爽亮好多，像是京剧里的西皮流水，要的就是这个爽利劲儿。

Xiaochang Chen's is a time-honored restaurant dating back to 1886. It is common to see diners form a long line to wait for its signature offer—*luzhu*. After eating *luzhu*, most old Beijingers order some pickled mustard Chinese cabbage to cleanse the palate.

Pickled mustard Chinese cabbage is a unique cold dish of Beijing cuisine, with a pungent taste comparable to that of *douzhir* or fried stinky tofu. For many outsiders or foreigners, pickled mustard Chinese cabbage is something they never want to try again. A single bite is enough to make one shed streams of tears and snivel. However, the dish remains popular among elderly Beijingers. Its ingredients are cheap and common: Chinese cabbage, yellow mustard, plus a little sugar and vinegar. The cold outer leaves of the Chinese cabbage, seasoned with spicy mustard, taste extremely refreshing. Due to its effect in removing grease and promoting digestion, the dish is often found alongside meaty ones on the winter dining table when the Chinese New Year draws near. The *luzhu* offered by Xiaochang Chen's Restaurant is greasy and savory. Eating it with pickled mustard Chinese cabbage will relieve pressure on the digestive system, and its refreshing effect feels like enjoying a wonderful performance of Peking Opera.

小腸陳
老字號誠信仁和

天子脚下，京城四方，什么才是“官府菜”？如何把自己的味蕾调整为“北京时间”？您可能先要从“八大楼”“八大居”和“四大顺”吃起。老字号有讲究，其中“楼”最大，既能请客宴饮也能办堂会。“居”规模要小一些，只能请客吃饭。招牌带“顺”的，清真馆子居多。

八大楼、八大居和四大顺——这一章都是当年皇亲国戚、北平文艺名流宴请之地。道道菜有讲儿。

If you want to experience Guanfu cuisine (literally "officials' cuisine," a culinary genre developed by officials and literati in old Beijing), the best choices are the famous time-honored restaurants collectively known as "Eight Lou," "Eight Ju," and "Four Shun." The "Eight Lou" restaurants are larger in scale and suitable for holding banquets and parties, the "Eight Ju" restaurants are smaller but also ideal to feast guests, and those whose names contain the character *shun* are usually Halal restaurants.

The famous "Eight Lou," "Eight Ju," and "Four Shun" are all restaurants where royals and celebrities in old-time Beijing held banquets.

第二章
Chapter 2

八大楼、八大居和四大顺

Famous Time-Honored Restaurants

01

东城区东直门内大街 5 号 · 5 Dongzhimennei Street, Dongcheng District

去东兴楼，尝一回“八大楼”之首的美食
Enjoy Delicacies at Dongxing Lou, Top of the "Eight Lou" Restaurants

东兴楼是正经的北京“八大楼”之首，不掺水分的“百年老店”——光绪二十八年（1902）开业的“中华老字号”。上大学没离开北京，有南方同学来了惊呼：“你们北京菜怎么什么都放酱油?”——北京的“官府菜”都是鲁菜打底。比如东兴楼的“招牌”葱烧海参、油焖大虾，讲究的就是“工艺复杂、食材高端”。

现在很多年轻人可能不理解，认为一盘“油焖大虾”有什么高端的，但这是北京官府菜的特殊历史造就的。不往远了说，马未都先生回忆徐邦达先生在新中国成立后，住在“东四”的时候，能一个礼拜吃一回“油焖大虾”，都是不得了的“大事”。那时候老北京人肚子里的油水少，所以一次体面的请客，肯定首选“东兴楼”，要让客人吃好。除了东直门店，其实您还可以来副中心通州店——有运河的水景儿，而且好停车。至于黑乎乎的“酱油”，没有酱油那还了得，那是鲁菜的灵魂。

Dongxing Lou, founded in 1902, is a time-honored restaurant, known as the top of the famous "Eight Lou" restaurants in Beijing. When I was studying in university in Beijing, one of my classmates from southern China once complained, "Why do almost all dishes of Beijing cuisine use soy sauce?" This is because Beijing's Guanfu cuisine originated from Shandong cuisine, which features "complicated culinary techniques and high-end ingredients." The most representative dishes are braised sea cucumber with scallion and prawns braised in oil from Dongxing lou.

Perhaps today's youngsters can't understand why prawns braised in oil was considered a high-end dish. This can be attributed to the unique history of Beijing's Guanfu cuisine. Ma Weidu, a famous Chinese antique collector, recalled that famous artist Xu Bangda would be excited to eat prawns braised in oil once a week when he dwelled in Beijing's Dongsi area. At that time, few Beijingers could afford eating meat very often. To entertain distinguished guests, they often chose Dongxing Lou. Alongside the Dongzhimen location, Dongxing Lou has another location in Tongzhou District, Beijing's Municipal Administrative Center. There, diners can enjoy the scenic views of the Grand Canal and find parking much easier. Soy sauce may make a dish less attractive in appearance, but it is regarded as the soul of Shandong cuisine.

東興樓
東來紫氣百年品質同作記
興來春風四方賓客共登樓

02

西城区前门西大街 2 号 · 2 Qianmen West Street, Xicheng District

去泰丰楼，体验一回“唱戏的腔，厨师的汤”
Taste the Chef's Soup Blending Diverse Flavors at Taifeng Lou

“唱戏的腔，厨师的汤”，虽然泰丰楼的创始人也是山东人，但汤做得地道，味道南北兼顾。皇城根下几朝古都，“你方唱罢我登场”，到了民国时期，孙中山、宋庆龄、蒋介石、袁世凯……您没听错，当时的北平，那些有头有脸的大人物宴请，都喜欢来泰丰楼，除了地段儿好——在前门方便，还有就是口味“南北兼顾”。

泰丰楼现在最有名的菜是“油炸双脆”——得提前一天预定，这是为了保持食材的新鲜。“双脆”，指的就是鸡胗和猪肚。按说这原料不是什么“山珍海味”，能成为“招牌”，主要是个功夫钱——成鲁菜的咸鲜口确实太困难了。略懂厨艺的人都懂，但凡内脏，基本多少要搭配点儿辣椒来炒，为的是去腥。能做咸鲜口儿的内脏，除了食材本身必须特别新鲜之外，“油爆”也需要大厨手艺：火要足、油温要特别高、速度要快，鸡胗和猪肚要做到脆而不生，稍微掌握不好火候就欠了或者过了，吃的就是一个手艺钱。会吃的“老饕”，到外边餐馆都愿意点这种“爆炒”，因为普通楼房里的天然气达不到这么足的火，爆不出来这个味儿。

The founder of Taifeng Lou also hailed from Shandong Province, but he excelled in cooking soup that blended flavors from both northern and southern China. Historically, the city of Beijing witnessed dynamic changes several times. During the Republic of China period (1912-1949), many celebrated figures such as Sun Yat-sen, Soong Ching Ling, Chiang Kai-shek, and Yuan Shikai preferred Taifeng Lou when entertaining guests, not only due to its prime location at Qianmen but also because the restaurant offered both northern and southern cuisines.

The most famous dish of Taifeng Lou is deep-fried chicken gizzards and pork tripe, which requires a reservation a day in advance. The dish's main ingredients, chicken gizzards and pork tripe, are neither rare nor expensive. Many wonder why it became the restaurant's signature dish. The secret lies in the difficulty in cooking this dish. Those familiar with cooking know that animal offal usually needs to be cooked with chili pepper to remove its pungent smell, so the finished dishes are often spicy. However, deep-fried chicken gizzards and pork tripe at Taifeng Lou boasts a fresh and salty flavor. To achieve it, the ingredients must be very fresh, and it also requires superb cooking techniques. The chef must deep-fry the ingredients in hot oil very quickly, so that the chicken gizzards and port tripe will be crispy. If the cooking temperature is inadequate or excessive, it won't be very crispy. Therefore, gourmets often order this dish at Taifeng Lou because it is so hard to cook at home as the natural gas stoves in most houses cannot reach the required temperature.

03

西城区前门西大街前门国帅宾馆
Qianmen Guoshuai Hotel, Qianmen West Street, Xicheng District

去致美楼，感受一次“御厨遗风”
Experience the Delicacies of the “Imperial Kitchen” at Zhimei Lou

作为“八大楼”之一，当年的致美楼红极一时，是道光年间就有的老字号，有御厨遗风，皇亲国戚、艺苑名流都曾是常客。致美楼地段儿也好，在大栅栏附近，挨着同仁堂，如果您正好是来前门，不妨顺便走进一回老字号。他家的招牌是“一鱼四吃”——“松鼠、糟溜、软炸、红烧”，味道依然是鲁菜的底子，见仁见智。不过，高峰期间急着赶路的朋友不建议来，厨师要一道一道地出菜。不要催菜……要是上菜慢了，您就当磨磨性子，安慰自己今天吃的不是菜，是“Beijing History（北京的历史）”。

Zhimei Lou is one of the “Eight Lou” restaurants. As a time-honored restaurant dating back to the reign of Emperor Daoguang (1821-1850) of the Qing Dynasty (1616-1911), it used to enjoy great popularity among royals and celebrities due to its culinary style modeled after that of the Imperial Kitchen. Its location next to the famous traditional pharmacy Tongrentang gives the restaurant an advantageous position in Dashilanr, a commercial street in Qianmen area. Tourists visiting Qianmen are suggested dining here. Its signature dish is One Fish Cooked in Four Ways—namely, deep-fried with sweet and sour sauce, stewed in fermented rice sauce, soft-fried, and braised in soy sauce. Naturally, this dish belongs to Shandong cuisine. Perhaps it doesn’t suit those in a hurry: it takes some time to prepare. So, be patient if you want to try it. After all, you’re tasting not only a dish, but also Beijing’s history.

致美樓飯莊
ZHIMEILOURESTAURANT

04

东城区王府井步行街 北京百货大楼 北楼六层 623 号 · No.623, F/6, North Tower, Beijing Department Store, Wangfujing Pedestrian Street, Dongcheng District

去萃华楼，掰一口“特色烤花卷”
Try Roasted Twisty Rolls at Cuihua Lou

惊不惊喜，意不意外，1940年开业的萃华楼同样是鲁菜！不知道是不是“八大楼”的老招牌会“劝退”年轻人（误认为消费太贵），有一次我路过长楹天街的“萃华楼”分店，发现楼牌宣传语已经变成了“一家小菜馆”。

不过，萃华楼真的属于名声在外、价格亲民的“八大楼”典范了，招牌菜还是经典鲁菜，九转大肠基本上桌桌必点。也有创新菜，比如特色烤花卷：有点儿像是黄油配烤面包，中西结合。但如果您属于“不爱肉、不爱甜”的瘦人饮食习惯，建议慎选萃华楼：他家点单率高的桃仁酱爆鸡丁、芝士南煎丸子、招牌酱汁鱼……多多少少或肉多或微有甜意。不过人生嘛，和朋友聚一聚，有点儿肉肉带点儿甜，才喜庆。

You may be surprised to find that Cuihua Lou, founded in 1940, is also a restaurant specializing in Shandong cuisine. The time-honored “Eight Lou” restaurants are not popular among today’s young people, many of whom mistake their dishes to be expensive. One day, I passed Cuihua Lou in the Paradise Walk Shopping Mall and found that its slogan had been changed to “A small eatery.”

Among the “Eight Lou” restaurants, Cuihua Lou is noted for its affordable offers. Of course its signature dish, braised intestines in brown sauce, also belongs to Shandong cuisine. It is a must-try for almost every diner in the restaurant. However, the restaurant also has innovative dishes such as roasted twisty rolls, which look like bread baked with butter. The dish combines Chinese and Western culinary styles. Perhaps Cuihua Lou isn’t an ideal choice for those who don’t like meaty, sweet food, since the best-sellers in the restaurant such as diced chicken and peach seeds in bean sauce, southern-style sautéed meatballs with cheese, and fish with the signature sauce are either a bit greasy or sweet. However, enjoying meaty and sweet delicacies with friends is all worth it.

萃華樓

05

西城区西四南大街 60 号 · 60 Xisi South Street, Xicheng District

去砂锅居，干瞪眼看对面小哥“包圆儿”砂锅白肉

Enjoy Pork and Vegetable Stewed in Chinese Clay Pot at Shaguo Ju

教您个北京菜小知识，凡是招牌写着某某“居”（比如砂锅居）的，说明当时规模不算太大——只能请客宴席，不能办堂会。西四的砂锅居是总店，高峰时段永远没地儿停车，慕名而来的都惦记这一口砂锅白肉。所谓“白肉”，就是用五花肉或者后臀尖儿，配上酸菜、粉丝，再下砂锅秘制，连汤带水，尤其适合北京“三九天”的冬天，吃完了脑门儿冒汗，从胃里暖到脚尖儿。要是赶上年轻力壮的小伙子吃砂锅白肉，再配一小瓶“牛二”（牛栏山二锅头），那真是一口白肉一口酒，四脖子汗流，一个人能“包圆儿”一锅，可以成仙了。

One thing you need to know about Beijing cuisine is that if a restaurant's name contains the word *Ju*, its size usually isn't very big. Such a restaurant is suited for treating several friends to dinner, but never for a large-scale celebration party. At peak hours, Shaguo Ju is always packed with diners, making it difficult to find parking nearby. Almost all diners try its signature dish—pork and vegetable stewed in Chinese clay pot. The dish is made of steak pork or pork haunch with pickled Chinese cabbage and glass noodles, stewed in a Chinese clay pot. The dish contains soup and is particularly suitable for dinner on a chilly winter day. One will sweat a bit when eating the dish and feel warm from head to toe. A young man can finish an entire pot alone while drinking a small bottle of *erguotou* (a representative of Beijing liquor), which can generate a feeling of euphoria.

老字号 砂锅居
砂锅居

06

西城区月坛南街甲 71 号 · A71 Yuetan South Street, Xicheng District

去同和居，享受一回当年鲁迅、齐白石都来的老字号

Enjoy Dinner at Tonghe Ju, a Restaurant Frequented by Lu Xun and Qi Baishi

同和居的招牌还是当年溥杰题的。如果您对溥杰印象不深，他有个哥哥您肯定听说过——末代皇帝溥仪。同和居作为“八大居”之一，也是当年鲁迅、齐白石、老舍……都来过的老字号。如今这里依然是鲁菜的底子，我们不再赘述，喜欢鲁菜的朋友可以把“八大楼”“八大居”的招牌菜都走一遍，味道轻重略微有所差异，各有所爱，看哪家更对您的口儿。

同和居还有一道点单率高的菜是三不沾，提醒头一次来的朋友注意的是，三不沾是一道甜品，点菜时不要把它当正菜。它的主料就是鸡蛋黄，味道偏甜。因为“不沾筷子、不沾盘子、不沾牙”，加上宫廷御膳的传说，才有了这道招牌菜。多说一句，很多外省朋友吃不惯北京菜，主要是因为点菜太看重“特色”了——如果你点的全是“特色”，很可能半桌子是“甜乎乎”的小吃，半桌子是“肉乎乎”的硬菜，连个清咸口的青菜都看不见，吃到胃里不一定好受。所以，招牌虽好，适量为珍。

The name of Tonghe Ju was inscribed by Pujie, a younger brother of China's last emperor Puyi. As one of the "Eight Ju" restaurants, Tonghe Ju was frequented by historical celebrities such as famous writers Lu Xun, Lao She and renowned painter Qi Baishi. Today, most dishes it offers remain Shandong cuisine. Fans of Shandong cuisine can try all the signature dishes of the "Eight Lou" and "Eight Ju" restaurants. Those dishes vary in style and taste, and one can definitely find his or her favorite.

The most popular food at Tonghe Ju is *sanbuzhan* (literally "Three Non-sticky"). First-time visitors need to know that it is a dessert instead of a main course. With egg yolks as the main ingredient, it tastes sweet and doesn't stick to the chopsticks, plate and teeth, hence its name.

Many outlanders are unaccustomed to Beijing cuisine. If they order just local signature dishes, they will find that half are sticky, sweet snacks and the other half are meaty. Those from the south who prefer a light flavor may feel uncomfortable due to the lack of vegetables. Remember, less is more when you order famous dishes in a Beijing restaurant.

07

西城区新街口南大街 178 号 · 178 Xinjiekou South Street, Xicheng District

去柳泉居，体验一回舌尖上的“北京市非遗”京菜技艺

Tickle Your Taste Buds with Liuquan Ju's Intangible Cultural Heritage Delicacies

柳泉居是有四百来年的京城老字号了，始于明代隆庆年间。作为“八大居”之一，柳泉居的拔丝莲子是京城独一份——别人家做不来。史上的柳泉居是靠着一口甜水井卖北京黄酒起家的老店，现在最受年轻人欢迎的反而是豆包儿——来这儿吃饭的基本都会打包几个，属于“白嘴儿”能干俩的好吃级别。

当然，不能只吃豆包儿，柳泉居的正餐还是鲁菜，尤其是芫爆散丹，基本上是每桌必点的招牌菜。“散丹”其实就是羊肚儿（胃部）的一部分，因为上面好多小圆疙瘩，像撒在上面的药丸一样，所以雅称“散丹”。“芫”指的是“芫荽”，即“香菜”。柳泉居的芫爆散丹，挑的都是粗壮鲜嫩的香菜梗，炒出香味儿还不会像香菜叶那样出汤。有机会您可以来体验一把，发朋友圈的时候记得屏蔽一下“讨厌香菜星人”的朋友们。

Established during the reign of Emperor Longqing (1567-1572) of the Ming Dynasty (1368-1644), Liuquan Ju is a 400-year-old restaurant in Beijing. As one of the "Eight Ju" restaurants, it is particularly famous for its unique offer—candied lotus seeds. Initially, Liuquan Ju was a shop specializing in Beijing-style rice wine made with sweet water from a well in its courtyard. Now, its most popular snack among young customers is steamed buns stuffed with sweetened bean paste. Almost all diners will take several buns back home after dinner.

Of course, the primary offering at Liuquan Ju is still Shandong cuisine. A must-try is sautéed mutton tripe with coriander. Mutton tripe comes from lamb stomach, and its surface has densely distributed small round pimples that look like pills. To make the dish, the cook chooses thick and fresh coriander stems rather than leaves to prevent excess juice. Find a chance to try this signature dish. Remember, don't share this experience with friends who dislike coriander.

柳泉居饭庄
天
柳泉居饭庄
柳泉居

08

西城区大栅栏商业街7号吉龙老栅栏商城2楼 · F/2, Jilong Laoshilanr Shopping Mall, 7 Dashilanr Commercial Street, Xicheng District

去东来顺，小料一定得咬牙要麻酱底儿
Choose Sesame Paste Dipping Sauce for Hotpot at Donglai Shun

东来顺属于北京人尽皆知的涮肉老字号了，分店极多，您要是嫌前门的店排队人多，可以搜搜其他分店，基本在北京各个区县遍地开花，能开起来的，味道都算不赖。

如何优雅地“装内行”吃一顿东来顺涮肉呢？有三个让人“高看您一眼”的方法——第一，清汤锅底；第二，不点饮料；第三，麻酱小料儿。“清汤锅底”很好理解，对于老北京铜锅涮肉，会吃的人都只点清汤锅底，甚至认为“只有清汤才不是对铜锅涮肉的亵渎”。“不点饮料”，就是要“白嘴儿吃”，因为无论您点了什么酸甜口儿饮料，都会对味蕾产生影响，不能最佳地赏鉴铜锅涮肉的原有鲜味儿。

“麻酱小料”，就是小料一定要咬牙选麻酱底儿——其他无论什么红油、海鲜油及各种创新酱料……在吃东来顺的“老派”眼里，都“不正经”。“老北京”对麻酱真的是有几近执念的执着，我小时候的老派涮肉没什么其他料：就是麻酱底儿，再根据自己喜好加点儿酱豆腐、韭菜花。当然了，现在东来顺为了迎合四面八方的客人，麻酱小料也有了与时俱进的“口感改良”，把鲜味儿再提一个高度。得嘞，还是得去吃一回不是？

Donglai Shun is a well-known hotpot restaurant in Beijing. The time-honored restaurant has many branches. If you don't want to wait for a seat in the busiest Qianmen store, you can find another in any district of Beijing. All locations are worth trying.

How does one enjoy hotpot like a local gourmet? First, choose clear broth. Second, don't order any beverage. Third, choose sesame paste dipping sauce. Beijing-style hotpot uses copper pots, and some even believe that "all broths except for clear will destroy a copper hotpot." Moreover, any beverage, regardless of the flavor, may prevent you from enjoying the authentic freshness of copper hotpot.

Diners are strongly recommended to choose sesame paste dipping sauce. All other dipping sauces such as chili oil, seafood sauce, and various innovative sauces are considered "unorthodox" by old-fashioned diners. Old Beijingers are obsessed with sesame paste. When I was a child, sesame paste dipping sauce was the only type available in Beijing-style hotpot restaurants, and diners could add some fermented bean curds and Chinese leek paste according to their own taste. Of course, because it welcomes diners from around the world, Donglai Shun now offers a rich variety of dipping sauces. Even the traditional sesame paste dipping sauce has been upgraded. Perhaps you can give it a try.

東来順
東来順

09

西城区南菜园街 12 号（大观园西门南侧）· 12 Nancaiyuan Street (south of the west entrance of the Grand View Garden), Xicheng District

去南来顺，感受一次地道南城味道
Experience the Authentic Delicacies of Southern Beijing at Nanlai Shun

电影《老炮儿》里有这样一段台词："六爷，没宣武区了，都合西城了……"一股老北京的忧伤弥漫而来。互联网的记忆好短，我在某潮流时尚App（应用程序）上刷到有人讲"你知道过去的宣武、崇文吗?"心想这不是常识吗？还有人问？后来一想，也改了十多年了。又一批长大的孩子，老宣武和南城文化的记忆已经淡得模糊得只剩个影儿了。

南来顺，菜如其名，从地理位置到菜品口味都是"南城味道"。菜品的口味很家常：从面茶、奶油炸糕、门钉肉饼到羊杂汤、炸松肉、清真的各色热菜都有，算不上哪一道的口味多"创新惊艳"、非要吃一口涕泗横流恨不得题块儿匾的程度，但就是觉得"吃着舒服、安心"。南菜园街店的位置，离着大观园不远。您要是春秋天来游个园，不妨顺便尝尝记忆中老宣武的"南城味道"。

In the movie *Mr. Six*, a character laments the tremendous changes of Beijing by saying sadly, "Mr. Six, Xuanwu District has been merged into Xicheng District." Someone once asked me whether I know anything about the former Xuanwu and Chongwen districts. My first thought was to ask why they wanted to know. Then, I realized that the two districts had been annulled for more than a decade. For the newest generation, memories of Xuanwu in southern old Beijing and its culture may be gone.

Just as the name suggests, Nanlai Shun (literally "Success in the South") stores are mostly located in southern Beijing and offer foods popular in the locale such as seasoned millet mush, fried butter cake, doornail meat pie, mutton offal soup, fried minced meat paste, and various kinds of Halal snacks. Most of these dishes are simple comfort foods. The Nanlai Shun store on Nancaiyuan Street is close to the Grand View Garden. If you visit the garden in spring or autumn, you should dine in the restaurant, which may awaken memories of the former Xuanwu District.

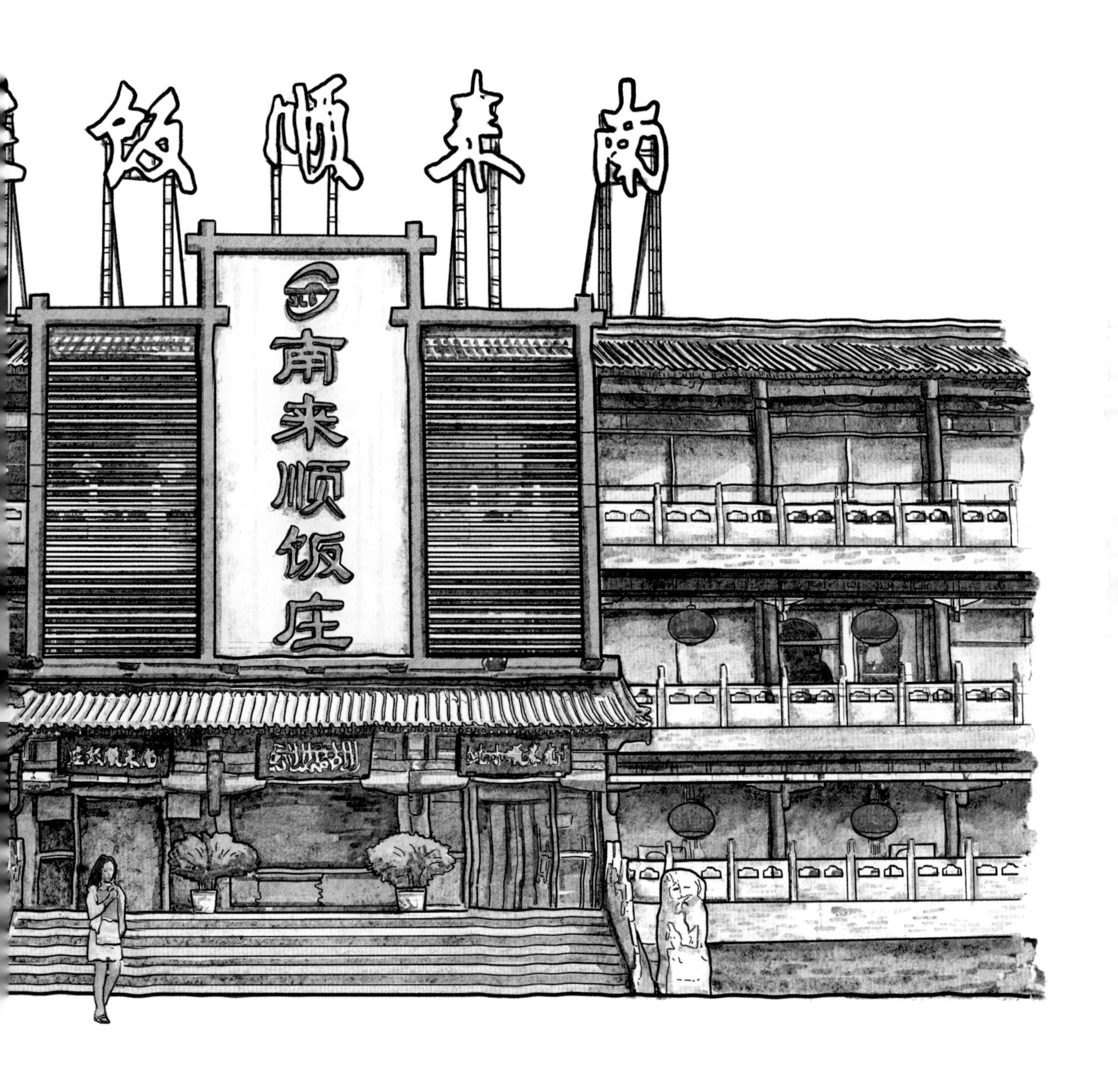
南来顺饭庄

10

西城区西长安街街道和平门北新华街 116 号
116 North Xinhua Street, Hepingmen, Xicheng District

去西来顺，点一份“马连良鸭子”
Taste Ma Lianliang Duck at Xilai Shun

一顿西来顺，半部民国史。当年西来顺的主厨褚祥出身厨师世家，早年在清宫御膳房当差学过西餐。后来民国了，褚祥就用“西菜中做、介入清真”的方式，独创了“清真西派菜”。

到西来顺，必点的一道菜就是“马连良鸭子”。您可能问，马连良不是京剧的名角儿吗？怎么给鸭子做代言了？这还是西来顺的一段“报恩菜”佳话。当年京剧“马派”当红，马先生来

用餐，赶上两伙人争包间，势力都不小，民国兵荒马乱，真打起来饭馆的桌椅板凳、陈设器皿损失就大了。马先生仗义出手巧妙化解了矛盾，褚祥为感谢马先生，就用鲁菜的香酥法，配以淮扬风味汤料，特制了香酥鸭，并亲自冠名“马连良鸭子”以示心意——腌入味的鸭子滋味十足，外酥里嫩，权当是“香酥鸭”升级版，口感错不了。

The history of Xilai Shun reflects half of the history of the Republic of China period. Its first chef, Chu Xiang, was born to a cook's family. He once worked in the Imperial Kitchen during the Qing Dynasty and later studied Western cuisine. During the Republic of China period, Chu invented the Western Halal style by combining Chinese and Western culinary techniques with Halal cuisine.

A must-try for diners visiting Xilai Shun is Ma Lianliang duck, a dish named after renowned Peking Opera artist Ma Lianliang. There is a story behind the origin of this dish: Once, when Ma Lianliang was dining at Xilai Shun, two groups of influential diners fell into a quarrel. Considering the social unrest during the Republic of China period, the restaurant could have suffered significant losses if the quarrel escalated into a fight. Thanks to mediation by Ma, the two groups eventually reconciled. To thank Ma for his help, chef Chu Xiang cooked a crispy duck using the culinary techniques of both Shandong and Huaiyang cuisines and named it Ma Lianliang duck. An upgraded version of ordinary crispy duck, the dish features salty duck which tastes crispy outside and tender inside.

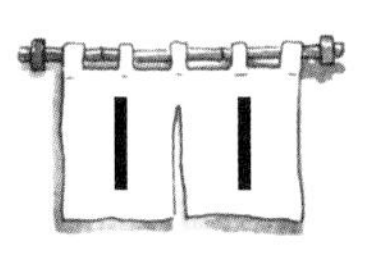

西城区黄寺大街 28 号 · 28 Huangsi Street, Xicheng District

去又一顺，皇亲国戚曾是座上宾
Dine at Youyi Shun like the Royals

又一顺同样是清真老字号。当时又一顺的贵客都是了不得的大人物——溥仪的弟弟溥杰等显贵皆曾是座上宾。他家的老招牌是醋熘木须。您别小看这道家常菜，好像是个清真馆子的菜单上都有，但真正能做好的人不多。

又一顺传统醋熘木须，一定要用磨裆肉——就是羊后腿臀尖下面，位于两腿裆相磨处的肉，肉质比较粗，肥多瘦少，边上稍有薄筋，腌制后用来做木须正好。既然是“醋熘”，这道菜放醋

也有讲究——要两次加醋，要的就是能吃到醋的馥郁香味儿，又不至于太酸。您要是喜欢地道的浓油赤酱咸口类清真菜，来这儿准没跑儿。

Youyi Shun is another time-honored Halal restaurant. Historically, it attracted many distinguished guests including Pujie, a younger brother of China's last emperor Puyi. The restaurant's signature dish is sautéed Moo Shu mutton with black vinegar. Though a common dish in Chinese Halal restaurants, it isn't easy to cook this dish well.

The chef at Youyi Shun uses mutton rump with a thin layer of tendon, which is gathered from between the two legs of a lamb and tastes fatty but chewy. He needs to add black vinegar twice in the cooking process so that the finished dish can maintain the fragrant smell of the black vinegar and won't taste too sour. Youyi Shun won't let you down if you like oily, salty Halal foods.

真正会吃的人，都好一口“老国营”。互联网化的年轻餐厅淡化了南北差异，只有在不变的“老国营”里，才保留了一个城市浓烈而倔强的“老派”味觉和方言记忆。够劲儿。

True gourmets prefer time-honored state-run restaurants. In the Internet age, modern restaurants often fuse northern and southern cuisines. Perhaps only state-run restaurants still retain the city's old-fashioned culinary style and memories.

第三章
Chapter 3

老国营 老味道

State-Run Restaurants and Groceries

01

东城区赵府街 67 号 · 67 Zhaofu Street, Dongcheng District

去赵府街副食店，拎着瓶子来打散装“二八酱”

Buy Bulk Mixed Paste at the Zhaofujie Grocery

据说，这是北京硕果仅存的一家老国营副食店了——旧鼓楼大街东侧的赵府街，清代属镶黄旗的地界儿，副食店门口唯一的装饰，就是玻璃上贴着俩极有年代感的大红字“国营”。一进去像是“穿越”到二十世纪七十年代，老柜台、散装麻酱，甚至连墙上海报贴的糖果的“糖”字、烟酒的“烟”字，现在四十岁以下的人可能都不认识（二十世纪七十年代的简化字，后来

又废除了）。

来这儿买副食的，可不仅仅是附近的老居民大爷大妈，现在乌泱泱排队的是慕名而来的年轻人——整书包的一瓶瓶囤“二八酱”。“二八酱”可能好多岁数小的朋友都没听过：就是两成芝麻酱、八成花生酱的“调和酱”。您可以自己带瓶子散打，也可以买成罐儿的。在计划经济时代，“二八酱”真是一个伟大发明：因为花生的成本比芝麻便宜，“二八酱”既取了芝麻酱的香味，又保持了花生酱的浓郁，成本还比纯芝麻酱便宜——散称十四块钱一斤。嘿！含嘴里五分钟香味儿不带散的，吃老北京火锅涮肉、早点抹馒头、夏天拌面条，就靠这口“二八酱”了。

要提醒一句的是，人家老国营不开网店，“二八酱”十四块钱一斤国营良心价。您要是不想自己跑腿儿，随便搜网上都有“网红代购”——九十八块钱一对儿，顺丰包邮。咱还是不惯那个不良风气，自己现场来打的酱，香啊。

This might be the only old state-run grocery store still existing in Beijing. Located on Zhaofu Street east of Gulou Street, once the territory of the Bordered Yellow Banner in old Beijing, the grocery features two red Chinese characters meaning "state-run" on its time-honored glass windows. Visitors may feel like traveling back to the 1970s: old-style counters, bulk sesame paste on sale, posters advertising candies and cigarettes... Those younger than 40 may be unable to read the earliest simplified traditional Chinese characters (adopted in the 1970s and later abolished) on the posters.

In addition to local elders, the grocery also attracts youngsters who come to buy Mixed Paste because it gained fame on the Internet. Many of the younger generation have no idea what Mixed Paste is. The paste is made of 20-percent sesame paste and 80-percent peanut paste. The store sells it both in bulk and bottled. In the era of the planned economy, Mixed Paste was a great invention: Due to the introduction of lower-cost peanut material, the bulk Mixed Paste is priced at 28 yuan per kilogram, much cheaper than pure sesame paste. However, its flavor is almost as fragrant as sesame butter. Taking a bite, you'll find the flavor can last for at least five minutes in the mouth. Mixed paste is an essential condiment for Beijing-style hotpot. It is also suited to mix with steamed buns and noodles.

It should be noted that the state-run grocery store doesn't have an online store. If you don't want to buy its famous Mixed Paste in person, you can ask others via the Internet to buy and mail it to you, but you will have to pay more—the price per kilogram is 98 yuan instead of 28. Let us just buy the paste in person ourselves—the paste tastes much more flavorful after putting in the hard work.

02

通州区南大街 12 号 · 12 South Street, Tongzhou District

去小楼饭店，一定得来一份烧鲇鱼
Try Catfish Braised in Soy Sauce at the Xiaolou Restaurant

来，先跟我念一个地名：南街。虽然现在的路牌和导航都变成了“南大街”，但您要到副中心问路遇到“老通县（通州在1997年之前，旧称通县。祖祖辈辈在这里扎根的老通州人，会自诩‘老通县’）”，人家可能下意识地一愣：“南大街？您说南街啊。您别往里开（车）了，紧东头儿（北京话：最东边的意思）是大高坡，汽车下不去。”

小楼饭店名声在外，光绪二十六年创建的清真老字号，在通州无人不知。小楼饭店的地理位置，非常符合通州二字的定位——四通八达，一条在乾隆年间被“切成十八个半截儿”的南街，看上去一条街上到处都是胡同口儿，到处都能过。建议自信人士开车来体验——像是“开盲盒”，在很多的半截儿胡同过五关斩六将后需要您“掉头，已为您重新规划路线”。

小楼饭店从菜系到服务都属于地道老国营——您点菜点多了，服务员大妈会劝您“打住”，绝对不让您花冤枉钱请客浪费。喝白开水自己去倒，装修的风格就是没装修。看菜单时，跟我念一个发音：“卷果（guo，轻声）”。这里基本上每桌必点的就是特色烧鲇鱼——挨着运河，老

事年间这里鱼多，但鲇鱼土腥味大，就把鱼去头尾留中段，切成块儿，用绿豆淀粉一裹，香油红烧，三炖三烤，急火文火反复过油（那油是用胡椒焙过的，能去土腥味儿）。最后，倒入辅料勾芡出锅。可以一尝，一尝乡愁双泪流。

With the help of road signs and a navigation app, finding South Street is easy. But if you ask an old native of Tongxian County (today's Tongzhou District) about South Street, he might be confused for a second and then reply, "Do you mean 'Nanjie'? It's at the easternmost end of this road. But, it is inaccessible by car because of an extremely high slope."

Built in 1900, the Xiaolou Restaurant is a time-honored Halal restaurant on South Street in Tongzhou District. Just as the name "Tongzhou" (a place extending in all directions) suggests, the street where the restaurant is located is part of an intricate road network with lanes extending to all directions. If you're very confident in your driving skills, you can drive there, but you might quickly get trapped in one of the many dead ends of the labyrinth.

Xiaolou is a state-run restaurant. If you order too much, the waiter will stop you for fear of waste. If you need water, serve yourself. And the restaurant's interior decoration is extremely simple. In addition to Yam and Date Rolls, another must-try is catfish braised in soy sauce. Close to the Grand Canal, Tongzhou abounds in fish. To remove the bilgy smell of the catfish, the cook cuts it into several pieces, wraps them with mung bean starch, deep-fries them in pepper oil, braises and bakes them repeatedly in high and then low heat, and finally add the sauce. The dish has the power to make diners feel nostalgic of their hometowns.

03

通州区新华东街 238 号 · 238 Xinhua East Street, Tongzhou District

去大顺斋，寻觅老国营的糖火烧
Buy Sugared Sesame Buns at the State-Run Dashun Zhai

明崇祯年间创建的大顺斋，也属于副中心大运河文化的一部分，老店址也在南街——南街北口儿往东拐一点儿，就到了运河漕运码头。来来往往运粮卸货或赶路，过去没有外卖保鲜盒，路上带得最多的是点心饽饽，尤其是大顺斋的糖火烧——既扛饿解馋，保质期又特别长，还不怕压坏了，随便往包袱里一放就行，不像是娇气的酥皮儿点心，颠簸二里路直接变点心渣了。

老国营的大顺斋糖火烧，看上去特不起眼儿——红糖色儿的火烧谁没见过啊。但人家老字号用料工艺都有讲儿。面是本地新麦子磨的；香油、麻酱的原料都是上等的白芝麻；火烧是死面还是发面的？都不全是。捏火烧时，要“两层死面”加“一层发面”叠起，上炉时候，饼铛内先文火抢脸儿一两分钟，然后上盘入炉，烘烤十五分钟，这才算大功告成。

现在都讲究“控糖”，可能是为了适应年轻人的口味，很多地方的糖火烧都没那么甜了。只有大顺斋以老国营的不变应万变，保留着它的甜——糖火烧就是得甜，不甜风味全无。然后我发现了老字号糖火烧的“最Fashion（时尚）的吃法”：原来它的灵魂伴侣是配浓苦的黑美式（咖啡）!

这就是大运河版的“马卡龙”。向晚意不适，糖火烧配热美式。

Dashun Zhai, established in the reign of Emperor Chongzhen (1628-1644) of the Ming Dynasty, is part of the Grand Canal culture in Tongzhou District. It was originally located on South Street. Turning east from the south end of the street gets you to the former freight terminal of the Grand Canal. In the past, it was packed with cargo and passengers. Back then, lacking food preservation devices, travelers often took steamed buns as food for the trip. Dashun Zhai's sugared sesame buns were popular due to their delicious taste and long shelf life. Moreover, such buns are pressure-resistant, making them easy to pack. Other pastries break into pieces after a kilometer of jolting.

The sugared sesame buns offered by the state-run Dashun Zhai seem ordinary, but the ingredients are carefully selected—flour made of newly harvested local wheat, sesame oil and paste made of top-quality white sesame seeds. The cook uses two layers of unleavened dough and one layer of leavened dough to make the buns, bakes them in the oven with low heat for one to two minutes, and then with high heat for 15 minutes.

Today, extra sugar is considered unhealthy. To meet the demands of young people in contemporary times, sugared sesame buns in many restaurants are not as sweet as they used to be. However, Dashun Zhai is an exception and strictly holds that sugared sesame buns must be sweet. I found by chance that the sugared sesame buns offered by the time-honored restaurant have a "soul mate": a thick and bitter Americano.

It is leisurely to eat a sugared sesame bun, nicknamed the "macaron of the Grand Canal," while sipping an Americano in the evening.

清真大順齋食品店

04

西城区前海东沿 14 号 · 14 Qianhai Dongyan, Xicheng District

去烤肉季（什刹海总店），体验一回“武吃”
Experience “Martial Dining” at the Shichahai Flagship Store of Kaorou Ji

如果您大冬天晚上在什刹海一带溜达，看到前面灯火通明一堆人在路边的板凳上坐着刷手机，估计就是烤肉季在排队——超火的老国营，永远排不上号的百年老字号。老北京烤肉分“文吃”和“武吃”。所谓“文吃”是指烤肉师傅将肉烤好，端给顾客直接食用；“武吃”是自助，顾客自己边烤边吃——这和满人传统有关，老事年间的“标准武吃架势”，需要全程围炉而立，一脚踩地，一脚踩在板凳上，讲究的主儿肩上还要挂一条毛巾擦汗，一手烧饼，一手羊肉，边烤边大口吃肉。

现在“武吃”基本失传了，烤肉季算是独一份。有外省朋友问，“武吃”会不会显得不够绅士、上不了台面？那您就大错特错了。当年在烤肉季“武吃”的爷，多是八旗贵胄。

好吃肉的朋友可以体验一把“武吃”——需要提前预约，有时候能赶上国家非遗传承人亲自烤。不过现在为了顾客安全，都是师傅烤完了帮您端上来。师傅要问您烤到什么程度，这里完全没有西餐牛排“几分熟”的“鄙视链”——您可以大大方方要全熟。肉可以分嫩、老、焦、糊，有人喜欢鲜嫩，老饕还专门要“焦香”，才显得老派高级。

If you stroll Shichahai Lake and see a group of people sitting on roadside stools while burying themselves in their smartphones outside Kaorou Ji, they are waiting for a seat in this century-old BBQ restaurant. A state-run restaurant, it is always packed with diners at peak hours. In Beijing, there are two ways to enjoy traditional BBQ: “literary dining” and “martial dining.” The “literary dining” refers to meat grilled by the cook in the kitchen and then served on the table, and “martial dining” refers to diners grilling the meat themselves—a tradition handed down from generation to generation among the Manchu people. In a standard “martial dining” process, the diners stand around the grill with one foot on the ground and another on a stool. Some hang a towel on the shoulder which they use to wipe away sweat as they enjoy grilled mutton and sesame buns.

Today, “martial dining” can only be found at Kaorou Ji. Some outsiders argue that “martial dining” is somehow uncouth. In fact, during the Qing Dynasty, “martial dining” was mostly practiced by Manchu noblemen from the Eight Banners.

You can experience “martial dining” with friends in the restaurant, but remember to make a reservation in advance. If you're lucky, you'll be served by the restaurant's chef, an inheritor of a unique BBQ technique that has been listed as national intangible heritage. For the sake of customer safety, the cook is responsible for grilling nowadays. After you place the order, the waiter will ask, “How much would you like your steak grilled?” Don't worry. No one would judge you if you ask for “well-done,” which might be considered “amateur” in a Western-style restaurant. In Kaorou Ji, you're free to decide how much the meat is grilled. Old-fashioned gourmets usually want it “well-done and crispy.”

烤肉季

锦芳
京糕
艾窝窝
豆汁儿
豆面糕
炸糕

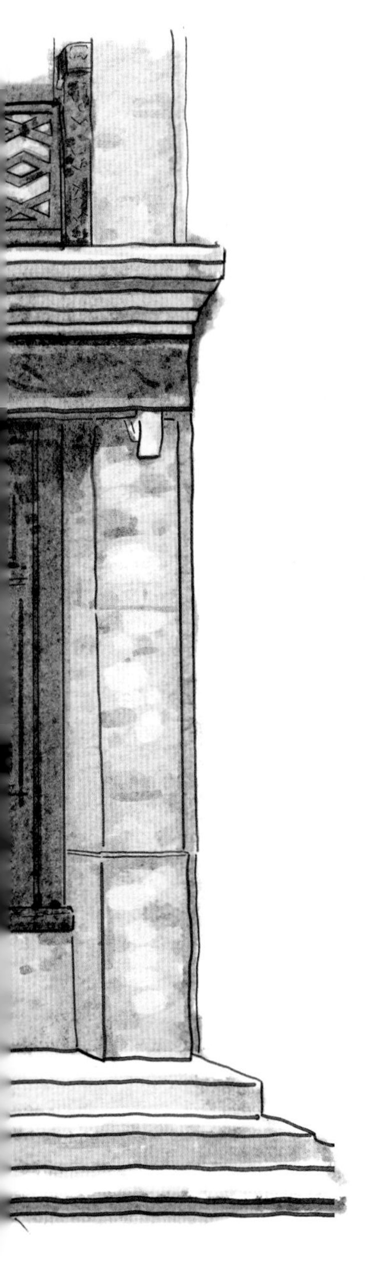

0 5

东城区磁器口路青年假日酒店 1 层（磁器口地铁站 D 西南口旁）
F/1, Youth Holiday Hotel (next to Exit D of Ciqikou Subway Station), Ciqikou Road, Dongcheng District

去锦芳小吃，尝一回老北京的什锦元宵
Taste Assorted *Yuanxiao* of Old Beijing at the Jinfang Snack Store

如果您想在元宵节前后，排队来买锦芳小吃的元宵，建议您穿厚点儿——可能要在寒风瑟瑟中排至少两个小时，才能买上他家的元宵。

锦芳小吃保持着老国营风貌——一个透明塑料袋子印着绿色大字“锦芳小吃”，各种馅儿一斤起卖。您要是小年轻尝鲜心态“这种馅儿我想来两个，那种馅儿我想来三个”，对不起，那您只能买一斤装的“什锦元宵”，配好了的各种馅料都有，不能挑。如果觉得来一回排队只买元宵太亏了，可以再加一份“奶油炸糕”，也是他家的招牌。

那您可能问了，为什么有那么多人排队？主要是老国营的元宵真材实料，绝对不会用糖精和杂七杂八的添加剂。

If you want to buy *yuanxiao* (glutinous rice balls) at the Jinfang Snack Store around the Lantern Festival, wear thick clothes. You'll have to wait in the long line for at least two hours in chilly winds.

Jinfang is a state-run snack store. All the *yuanxiao* it sells is packed in plastic bags with green Chinese characters translated as "Jinfang Snack," and the minimum for each kind is 500 grams. If you want to try several kinds but don't want to buy too much, assorted *yuanxiao* (several kinds mixed together) is the only choice. If you don't want to wait two hours just for a bag of *yuanxiao*, you can also buy the store's other signature offer: fried cream cakes.

Some wonder why the store's *yuanxiao* is so popular. The reason is the top-quality ingredients it uses to make *yuanxiao* and lack of any unhealthy additives. The state-run snack store prioritizes product quality over fancy packages.

06

西城区大栅栏街 30 号 · 30 Dashilanr Street, Xicheng District

去新成炸酱面馆，数数一碗炸酱面里有多少块儿肉丁儿

Count Diced Pork Pieces in a Bowl of Noodles with Soybean Paste at the Xincheng Noodle Restaurant

新成削面馆，顾名思义，当然是“新中国成立后的削面馆”，承载了一代老北京的老国营记忆。之前就是老辈儿人家常便饭吃碗刀削面的地儿——他家的浇头是传统的老方子，味道没有变过。后来改革，削面馆变成了新成炸酱面馆，但是地理位置没变，老店还是位于前门大栅栏，面还是那碗面，不过“面孔”变了——慕名而来的游客和小年轻居多。炸酱面这个事儿，没有“最正宗”和“最不正宗”：您路上拉十个老北京问问，谁家炸酱面最地道，肯定能得出十一个以上答案：全说自己家的最地道。但新城炸酱面的肉丁儿给得特足——老国营风范延续。

The noodle restaurant's name "Xincheng" (literally "newly founded") celebrates the founding of the People's Republic of China in 1949. It is a state-run restaurant carrying the memories of old-generation Beijingers. In the past, many local residents liked to eat a bowl of sliced noodles as dinner in the restaurant. The flavor of the noodle condiments it offers has remained unchanged for decades, but the signature offer changed from sliced noodles to noodles with soybean paste. Located on Dashilanr Street at Qianmen, a traditional commercial area in Beijing, the restaurant now attracts countless tourists and youngsters. Where can one find authentic noodles with soybean paste in Beijing? Perhaps everyone has an answer. The noodles offered by Xincheng feature abundant diced pork—state-run restaurants including Xincheng consider honesty a lifeline.

大栅栏总店
成炸酱面馆
CHENG ZHAJIANG MIANGUAN
30
地道北京
新成始于一九

07

东城区鲜鱼口街 81-83 号 · 81-83 Xianyukou Street, Dongcheng District

去天兴居，体验一把“喝炒肝儿”的老规矩
Eat *Chaogan* the Traditional Way at Tianxing Ju

天兴居也延续了“老国营”的气息。最地道的就是这一口炒肝儿——他家的炒肝儿除了“肝”，不加“心”“肺”，只留大肠，独门老手艺处理得特别干净。怎么看一碗北京炒肝儿好不好？芡要浓稠，碗里能立住勺子不倒才能给客人往桌上端。不过老事年间的炒肝儿不叫“吃”，叫“喝炒肝”。喝炒肝的规矩也大，不能用勺儿，得转圈吸溜才是老派的“灵魂吃法”——“炒肝儿不溜边儿，白在世上颠儿”嘛。

Tianxing Ju inherited the business style of state-run restaurants. *Chaogan* (fried liver) is its signature offer. Alongside pork liver, the *chaogan* it offers also contains intestines which are carefully processed with a unique method. How does one judge the quality of *chaogan*? First, a bowl of high-quality *chaogan* has thick soup. When you insert a tablespoon in the soup, it shouldn't fall. However, authentic gourmets drink *chaogan* directly without using a tablespoon. The traditional way is to sip *chaogan* along the edge of the bowl. A saying goes that "one's life is incomplete without siping *chaogan* along the edge of the bowl."

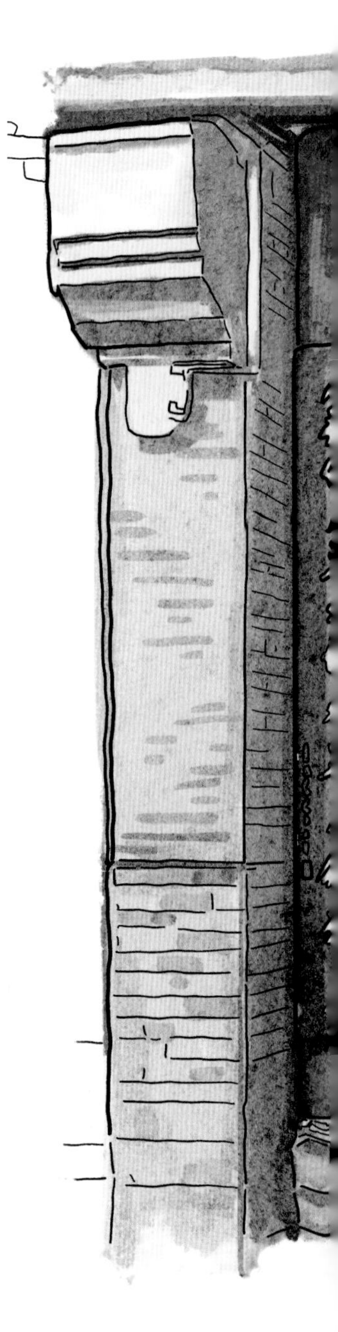

天興居

08

丰台区北大街南里5号 · 5 Beidajie Nanli, Fengtai District

去丰春楼饭庄，尝一口老国营的焦熘丸子
Try Deep-Fried Meatballs at the State-Run Fengchunlou Restaurant

一个简单低调的门脸儿，写着“老店始于1956（年）”，这就是丰春楼了，服务员以“北京大妈”为主，地道的北京家常菜，地道的北京话。您要是特意来北京想吃一顿特色菜，可能觉得这馆子没什么意思：既不在老字号“打卡”范围内，也没有“小清新”新派京菜装修讲究又唱曲儿又挂匾的“拔草”功能。来这里的，都是老北京，想在家常环境下，吃一口小时候的家

常菜。这就像是北京小伙儿谈恋爱，第一次约会可能约个姑娘喜欢的、“仙儿里仙儿气”的餐厅。什么时候来丰春楼了，说明关系近了：你看，这就是我小时候吃的焦熘丸子、滑熘里脊……嗯，你看过了二十年，还是这个味儿。

Fengchun Lou features a low-profile shopfront with a signboard carrying its name and the words "Since 1956." Most of its waiters are middle-aged local women speaking authentic Beijing dialect, and it offers ordinary dishes. Fengchun Lou is neither as famous as time-honored restaurants nor does it look like instagrammable modern Beijing-style eateries. Most diners come to re-experience flavors from their childhood. For a first date, a young man will likely take the date to a fancy restaurant. After the couple gets acquainted, he may finally take his girlfriend to Fengchun Lou and show her his favorite dishes from childhood. "They still taste the same two decades later."

09

西城区西长安街街道景山西街北海公园东门内
Inside the east gate of Beihai Park, Jingshan West Street, Xicheng District

去北海公园仿膳饭庄，尝一回真正“宫里的菜”

Taste Authentic Royal Cuisine at the Fangshan Restaurant in Beihai Park

您先听仿膳饭庄的地理位置——位于北海公园琼岛漪澜堂、道宁斋等一组乾隆年间兴建的古建筑群中，获国家二级企业称号，定点涉外餐馆，新中国成立后的很多重要中外领导人都在这儿吃过饭。“仿膳饭庄”在1925年创办，当时叫“仿膳茶社”——所谓“仿膳”，顾名思义就是仿照御膳房的菜，而且当时以茶点为主。那为什么叫“仿膳”不叫御膳呢？这是历史年代原因，那时候“推翻帝制”还没多久，怕叫“御膳”惹麻烦，但皇上没了，御膳房的伙计们也得重新谋出路吃饭不是？于是真的“御膳”反而叫“仿膳”了。

听到这儿是不是心里一咯噔：这地方是咱普通工薪阶层能去吃的吗？还真可以。仿膳饭庄平时也接待散座儿，人均5000块的满汉全席能做，人均200块左右的大众消费也接待。反正，不多说了，这辈子怎么也得吃一回啊。

A national class-II enterprise and foreign-related restaurant, the Fangshan Restaurant stands alongside a group of ancient buildings dating back to the reign of Emperor Qianlong (1736-1795) of the Qing Dynasty such as the Qilan Hall and Daoning Studio on Jade Islet in Beihai Park. After the founding of the People's Republic of China in 1949, the restaurant received many Chinese and foreign leaders. Founded in 1925, the Fangshan Restaurant was formerly known as the Fangshan Teahouse. "Fangshan" means "modeled after the cuisine of the Imperial Kitchen." Initially, it mainly offered tea and snacks. So, why wasn't it named Yushan (Imperial Kitchen) directly? Back then, the imperial regime had just been overthrown, so the restaurant was named Fangshan to avoid unnecessary troubles although many of its employees came from the Imperial Kitchen.

Many worry that ordinary people cannot afford to eat at the Fangshan Restaurant. In fact, it offers not only the Manchu-Han Imperial Feast which costs each diner 5,000 yuan, but also ordinary dishes. Nevertheless, the Fangshan Restaurant is a place worth trying in your lifetime.

仿膳
宫廷風味

东城区东四北大街48号 · 48 Dongsi North Street, Dongcheng District

去百年义利，找一口儿时的维生素面包
Recall Your Childhood by Tasting Vitamin Bread at the Century-Old Yili Bakery

每个北京小孩儿的记忆里都有义利面包厂的“维生素面包”。中华老字号，良心老国营。别小看这家面包厂，历次国家重大庆典活动中都是首选产品：北京奥运会、亚运会、阅兵式、国庆节、“全国人大”、“全国政协”……瞧瞧人家义利面包的“客户名单”，真是六块钱良心定价的果子面包后面，藏着共和国记忆。

“义利维生素面包”再次火出圈，是因为“逼死现代设计师”的设计颜色：对对，要的就是这种“大黄大蓝大粉”都用上的配色，一个都不能落，离着三米就能从货架上认出来有年代感的包装。跟你们说，你们那些“小清新”的爱马仕橙、香奈儿黑、蒂芙尼蓝的配色，都底气不足。人家真正的老字号，就得这么任性可爱地传承包装。你看我都过三十岁了，“义利维生素面包”的包装还是小时候的样子。看着它，就觉得还能回到小时候坐在妈妈自行车后座、偷偷咬一口“维生素面包”的童年似的。

Most Beijingers remember Yili's vitamin bread from childhood. A time-honored brand and state-run bakery, Yili has provided bread for banquets on many important occasions including the Beijing Olympics, Asian Games, National Day military parades and sessions of the National People's Congress, and the National Committee of the Chinese People's Political Consultative Conference. Behind its signature product, dried fruit bread priced at only 6 yuan each, are the memories of the People's Republic of China.

The recent popularity of its vitamin bread is attributed to the packaging fusing brightly yellow, blue and pink colors seldom used by contemporary designers, who prefer so-called "Hermès orange," "CHANEL black," and "Tiffany blue." However, this seemingly outdated package makes the bread stand out on the shelf. The courage of Yili to maintain this package design for more than three decades stems from its long-standing brand reputation. It awakens childhood memories of stealthily taking a bite of new Yili vitamin bread on the backseat of your mother's bicycle.

百年義利
CENTURY OF YILI

门头沟区新桥大街49号 · 49 Xinqiao Street, Mentougou District

去新新包子铺，尝一回“两元一个，只卖一种馅儿”的霸气包子
Buy Steamed Buns Stuffed with Pork and Scallion at Xinxin

如果您关掉手机随便在北京的街上转，如何通过吃货的第一直觉，判断一家店会不会好吃呢？新新包子铺基本就符合“老饕探店”的全部潜质：门脸简单，排队巨长，食客以本地的大爷大妈为主；再走进店面，发现菜单越简单，好吃的可能性越大——比如人家新新包子铺，几十年了，包子只卖一种馅儿：猪肉大葱。粥也只有一种：豆儿粥。价格非常美丽：包子40块钱一斤，一共20个。如果这时候您通过店里的装修，感受到一股“老国营”的浓烈气息，墙上的招牌字体就是“没字体”，装修就是“没装修”，连餐具都是二十世纪九十年代早餐店的“过气”铝勺儿，那就更没跑儿了，赶紧跟小伙伴“分工排队+占座”，一定得尝一回。这时候你就应该瑟瑟发抖地问一句前面排队的年轻人：“这里能用手机支付吧？”完了，脑子全被包子的香味占领，已经在幻想咬一口的味道了，智商降到冰点，怎么也想不起来“现金”怎么说，蹦出一句：“我没带……实体钱。”

How does one judge a restaurant's offers with instinct only, without using the Internet? Xinxin meets all the expectations for an old eatery: simple decorations, and a long line of local customers. After you enter the restaurant, you'll be stunned by its simple menu. For decades, the restaurant offered only one kind of steamed buns—stuffed with pork and scallion—and only one kind of porridge, bean congee. The steamed buns it offers are cheap at only 2 yuan each. The interior decoration is also the simple style of many old state-run restaurants, and even the tableware it uses are the outdated aluminum pieces often found in restaurants in the 1990s. If you come with your friend, you are suggested to divide the "labor" and have someone save a seat while the other orders food at the counter. But don't worry, mobile payments are accepted since few carry cash these days in China.

子铺

12

西城区西四北大街 24 号 · 24 Xisi North Street, Xicheng District

去杏园餐厅，吃一次“不带滤镜”的老北京糖醋里脊

Enjoy Authentic Beijing-Style Sautéed Sweet and Sour Pork Tenderloin at the Xingyuan Restaurant

在北京吃顿饭，经常出现“士别三日，当刮目相待”的情况：曾经觉得不起眼儿的老牌子餐厅，再认识它的时候已经在新闻上——莫名其妙成网红店了。老国营的杏园餐厅，现在火爆程度也是网红级别。之前只不过是觉得店里人多，现在高峰时候人多得不仅没位置，都快下不去脚儿了。人多了评论多了，就有慕名而来的小年轻觉得“报吃（“不好吃”在北京口语中的连读）”，没有传说和想象中的惊艳。但这里要为“老国营”们发个声：人均消费不到60块，北京家常菜老馆子，不能拿人均600块的“味觉盛宴”去苛求。至于“没吃过杏园餐厅，都不好意思说自己是老北京”的过誉评论，其实也有点儿过了：杏园餐厅主打的是山西刀削面，它从根儿上就不是北京菜，只能算是在北京历史比较久、融入了老北京口味、在附近许多老北京居民心中占了重要位置、有着许多人童年记忆的老国营餐厅。如果恰好路过，绝对值得来尝一尝。这是一家原本就“不带滤镜”的老北京平民餐厅，糖醋里脊也是“不带滤镜”的味道，又甜又咸，吃完了对男朋友一抹嘴：“皇上您可以去结账了，臣妾吃完了还要去搬砖。”

“One should look different even after only three days of separation,” goes a proverb. Indeed, a formerly unnoticeable restaurant may become Internet-celebrated for no reason in a flash. An old state-run restaurant, Xingyuan, is now popular on the Internet. At peak hours, it is always crowded with diners. Attracted by the flood of favored comments on the online catering platforms, countless young people come to experience the restaurant in person, but many are disappointed that its offers are not as delicious as the buzz indicated. However, the prices are highly affordable—dinners average around 60 yuan a head. So, don’t go expecting a MICHELIN Star.

People also say that “one cannot call himself a gourmet of Beijing cuisine without visiting the Xingyuan Restaurant.” But actually, the restaurant’s signature offer, sliced noodles, is from Shanxi cuisine. So, in essence, Xingyuan isn’t a Beijing-style restaurant. The reason it occupies a significant position in the minds of old-generation Beijingers is that this time-honored state-run restaurant carries childhood memories. It remains an ideal option if you happen to pass it at dinnertime. Its sautéed sweet and sour pork tenderloin features authentic Beijing flavor: sweet and salty. The restaurant is for everyday Beijing eating.

杏园餐厅
JUDEHUATIAN

这一章属于闻名中外的“北京招牌菜”系列——烤鸭、炸酱面、御膳饽饽……这些菜比起豆汁儿、焦圈儿，更能适应中西、南北的口味调和。

见过老外受不了芥末墩儿的，见过南方姑娘吃不惯卤煮的，但这一章的“北京招牌菜”，算是人见人爱。

This chapter focuses on signature dishes of Beijing cuisine that are known both at home and abroad such as roast duck, noodles with soybean paste, and royal steamed buns. Contrasting local snacks such as *douzhir* and fried crispy donuts, the dishes recommended here combine the flavors of different regions.

Perhaps pickled mustard Chinese cabbage or *luzhu* can never be accepted by outsiders, but the signature Beijing foods mentioned in this chapter cater to the tastes of almost everyone.

第四章
Chapter 4

中外闻名北京味儿

Beijing Flavors with Global Fame

LIQUN ROAST DUCK
RESTAURANT
福

东城区北翔凤胡同 11 号 · 11 North Xiangfeng Hutong, Dongcheng District

去利群烤鸭店，尝一尝胡同儿四合院里的北京果木烤鸭

Taste Beijing Roast Duck at the Liqun Restaurant Hidden in an Alley

知道利群烤鸭店的外国人可能要比中国人多——很多明星、驻华大使都特意来这儿，吃完了发“油管”和“脸书”晒完图，才算是来过北京。在鸭子的做法上，利群烤鸭和全聚德属于同一派系——当年全聚德的张利群师傅出来自立门户的烤鸭店——地道的挂炉果木烤鸭。前门北翔凤胡同，老老旧旧的四合院，里边桌子椅子的年龄可能都比我大了。

优点是这里确实是“打卡圣地”，毕竟那么多名人和大使都来过，而且菜单双语，地理位置又好，带外国朋友逛完前门直接胡同和北京烤鸭一趟行程搞定。缺点也非常明显，胡同不好停车。为避免剐蹭或者吃半截儿去给人挪车，最好先找好停车场，步行前往。怎么着？您是不是很好奇，这得是什么神仙味道的北京烤鸭，才这么傲娇还能这么火爆啊？要不……咱也去尝一回？

Maybe the Liqun Roast Duck Restaurant is more popular with foreigners than Chinese. During visits to Beijing, many foreign celebrities and diplomats like to dine here and then post selfies on YouTube or Facebook. Its duck roasting method is the same as Quanjude, another time-honored roast duck brand, both using fruit tree branches as fuel. In fact, the founder of Liqun, Zhang Liqun, once worked at Quanjude. The restaurant is tucked away in an old courtyard in North Xiangfeng Hutong at Qianmen. Perhaps every table or chair there is older than me.

The restaurant is worth trying. After all, many celebrities and foreign ambassadors have dined there. Its menu is bilingual. One can accompany a foreign friend to visit Beijing's *hutongs* and taste roast duck at the same time. Of course, it has shortcomings: it is difficult to find parking in the lane, and those who drive should park in a nearby parking lot and then walk to the restaurant. Despite so many shortcomings, the restaurant remains popular, evidencing the deliciousness of its offers.

02

西城区护国寺街 85 号 · 85 Huguosi Street, Xicheng District

去富华斋饽饽铺，尝一尝《红楼梦》里的宫廷饽饽

Taste Royal Steamed Buns Depicted in *Dream of the Red Chamber* at Fuhuazhai

富华斋饽饽铺有多火呢？在店里堂食是件充满“罪恶感”的事：因为排队的人太多，护国寺街寸土寸金，店面小，你刚咬两口，可能发现桌子旁边站着拿着盘子的妈妈带着小朋友，眼巴巴地看着您座位用没用完；或者“网红”小姐姐已经调好了美颜滤镜，就等您用完膳了能腾出个位置——人家直播预告都发出去了。

如果想带朋友来尝尝真正的北京味道，建议来几乎零差评的“宝藏饽饽铺”——当年宫廷御厨的后人王希富先生传下来的——我第一次尝的时候惊为天人，真的复原出了《红楼梦》里的瓜仁油松月饼！饽饽在方言里是“点心”的意思。一个小门脸全靠口碑相传，生意好到您都不好意思要壶乾隆“三清茶”配点心堂食听个古曲。如果您有特殊需求，这里的点心可以给您切好了配“刀叉”——中西结合。没错没错，就装在清宫审美的中国盘子里。透过大厅的玻璃窗能看到内厨在揉面，别小看这些点心揉面工！大有海外留学的名校硕士回来，心甘情愿来这儿学御厨饽饽手艺“揉面”的。哎，年轻人海外看了一圈儿世界，归来发现还是宫廷御厨的“揉面”高级。给祖国文化点个赞。

How popular is the Fuhuazhai Pastry Restaurant? Diners often get guilted into eating fast because so many people are waiting for seats. The restaurant has limited space. Before you finish your dinner, you may awkwardly find a mother and child holding a tray of food eagerly anticipating your departure or an online influencer preparing for a livestream next to you.

If you want to take your friends to taste authentic Beijing desserts, Fuhuazhai is the best choice and has almost zero negative comments online. It was founded by Wang Xifu, a descendant of the chef of the Imperial Kitchen in the Qing Dynasty. I was shocked when I first tasted the pine nut mooncake from the restaurant, which is identical to what is described in the classical novel *Dream of the Red Chamber*. The pastry is called *Bobo* in local dialect. Despite its small size, the restaurant's reputation has spread by word of mouth, leading to its astonishing popularity today. For foreign customers, the restaurant thoughtfully prepares knives and forks in traditional packages of the Qing aesthetic flavor—a combination of Chinese and Western styles.

Through the glass window in the restaurant, one can see cooks kneading dough in the kitchen. Don't underestimate these cooks. Some graduated with master's degrees from prestigious universities abroad and came to learn the pastry culinary techniques handed down from the Imperial Kitchen of the Qing Dynasty. After traveling around the world, those young people eventually realized the importance of carrying forward the cultural heritage of their motherland.

03

东城区王府井大街 301 号王府商厦 6 层 · F/6, Wangfu Commercial Building, 301 Wangfujing Street, Dongcheng District

去大董，体验一回“黑珍珠三钻”的北京烤鸭
Eat Roast Duck at Dadong, a "Once in a Lifetime" Restaurant

如果您的朋友或客户请您吃大董，不一定代表他很懂老北京烤鸭，但代表他一定很尊重您，您在对方心里是贵客——大董是新派烤鸭，“黑珍珠三钻”，意思是“一生必吃一次的餐厅”，人均消费属于“贵”但不“奢”的程度，商务宴请刚刚好。大董的“摆盘法”和“菜单”都是艺术：一盘儿最普通的宫保鸡丁都能给您摆出花儿来，端上来有面儿，适合请客。

很多人好奇，北京老字号烤鸭和大董新派烤鸭的区别，如果用一句话解释，就是老字号要的是“地道”，但大董的“新派”赢在“调和”——用艺术调性的方式，让南来北往的朋友都能接受。比如，老字号传统的北京烤鸭“油大”，在过去缺油少盐的年代，吃着香。但现在大家生活条件好了，很多人不愿意吃那么油的正餐，大董的招牌烤鸭“酥不腻”就更适合现代新派口味。另外，如果刚毕业的小年轻和学生党觉得大董太贵，可以去价格更亲民的小大董。

If your friend or business partner invites you to eat Beijing roast duck at the Dadong Restaurant, that does not necessarily mean that he is a gourmet, but reflects that he must consider you a distinguished guest. A modern-style roast duck restaurant, Dadong has been rated with "three diamonds" in the Black Pearl Restaurant Guide, the Chinese version of the MICHELIN Guide, which means "a restaurant that one must visit once in a lifetime." Of course, the food is expensive, but not unaffordable. The restaurant is ideal for business banquets. Dadong is noted for its art of plating—even a common dish like Kung Pao chicken can look extremely attractive there. So, it is an ideal place to treat distinguished guests.

Many may be curious about the difference between traditional and modern styles of Beijing roast duck. Simply speaking, the traditional roast duck offered by time-honored restaurants is renowned for its "orthodox flavor," while the modern roast duck offered by Dadong features a combined style with artistic touches, which caters to the taste of people from everywhere. For instance, the roast duck of time-honored brands is a bit greasy, which was popular in the past era when people barely had enough to eat. But today, with the continuous improvement of living standards, many prefer a healthier diet with less oil. In this context, Dadong's modern-style roast duck has become more popular. Students and recent graduates perhaps still cannot afford eating at Dadong. However, they can eat the same roast duck from its bargain brand, "Taste of Dadong," which is much cheaper.

大董烤鸭店
DADONG ROAST DUCK RESTAURANT
6F

04

西城区廊房头条21号院W9楼4层01号
No. 01, F/4, Building W9, Compound 21, Langfang Toutiao, Xicheng District

去“提督”，品尝烤鸭配西式鱼子酱
Experience Roast Duck with Caviar at TIDU

这也是一家“黑珍珠”入围餐厅。之所以把“提督”放到第四章中外融合系列，主要原因在于它作为一家闻名北京的知名“中餐厅”，很多只好传统中餐的长辈，来这里可能看不懂菜单搭配：对不起，这不就是北京“糖油饼”吗，配“樱桃清酒鹅肝”是什么组合？北京烤鸭和西式鱼子酱融合，做成“鱼子酱脆皮烤鸭”是啥味儿？请告诉我“莫吉托”和京葱爆羊肉能一起点吗？对，提督的特色，就是最传统的中餐，融合最国际化的“料理”。

“提督”嘛，古代一品封疆大吏称为“提督”，守护一方百姓、四海升平。当然了，提督也有地道京菜，同样是做烤鸭，提督的“五味三吃烤鸭”算是独创一绝。如果觉得“京兆尹”的档次过高、小年轻的钱包受不住，可以来吃个“提督”，人均消费要划算得多，也算是顿大餐了。

TIDU is also featured in the Black Pearl Restaurant Guide. Though a Chinese-style restaurant, older-generation diners may be a bit confused by its menu. Deep-fried pancake with cherry foie gras and sake, beijing roast duck with caviar, sautéed sliced mutton with scallion, and mojito might seem like “crimes” to them. Uniquely wild pairings is the theme of TIDU—blending traditional Chinese cuisine with international culinary styles.

In Chinese, TIDU means “provincial commanders-in-chief,” who were responsible for defending regional security and maintaining peace for local residents in ancient times. In addition to authentic Beijing dishes, TIDU also offers unique dishes such as Five-flavored Roast Duck Eaten in Three Ways. Moreover, compared to many high-end restaurants in Beijing, TIDU is less expensive, and even young people with humble salaries can afford it.

05

东城区方砖厂69号炸酱面（方砖厂胡同店）· 69 Fangzhuanchang Hutong, Dongcheng District

去方砖厂69号，尝一尝米其林一星的“北京炸酱面”

Enjoy Beijing-Style Noodles with Soybean Paste at MICHELIN Star-Rated Restaurant Fangzhuanchang No. 69

方砖厂69号的一碗炸酱面，都卖成了米其林一星，估计很多老北京听了，可能又“服气”又“不服气”。“不服气”的理由很简单，炸酱面不像是烤鸭，必须去饭馆吃，大家的品评有个高低上下。炸酱面的面和酱，老北京家家户户都能做，面的粗细、酱的浓稠、甜咸的配比、油多油少、面码儿放什么菜，家家户户都有自己的口味，都觉得自己家的“最正宗”。但是“服气”的理由是：人家真把老北京炸酱面推广成了“米其林一星”。

“方砖厂”的炸酱现在也能零售邮购。之前有外省来北京安家的朋友，跟我说他打算去订两罐“方砖厂”炸酱。我没吱声。然后另一个南方“胃软”的朋友说：“方砖厂的面真的又粗又硬！”我一机灵：“哎！这个正宗！这个炸酱面的手工面条，就得又粗又硬，裹上炸酱才有嚼劲儿。那咱们可以去吃吃！”这基本是最良心实惠的米其林一星了，怎么也得去捧个场是不？

Fangzhuanchang No. 69 is a noodle restaurant rated “one star” in the MICHELIN Guide. Many may feel the reputation is unfounded. Unlike roast duck, almost every household in Beijing can cook noodles with soybean paste at home, each has a unique recipe that they consider “most authentic.” However, none but Fangzhuanchang No. 69 has made the noodles into the MICHELIN Guide.

The restaurant now sells its homemade soybean paste online. A friend who moved to Beijing from elsewhere told me that he was going to order two cans of soybean paste from Fangzhuanchang No. 69. Then another friend from southern China who likes soft foods complained that the restaurant’s noodles are “too thick and hard.” “Hey, this is authentic!” I explained. “The handmade noodles taste chewy when mixed with soybean paste. Let’s give it a shot!” Perhaps Fangzhuanchang No. 69 is the most affordable restaurant in the MICHELIN Guide, and it is definitely worth a visit.

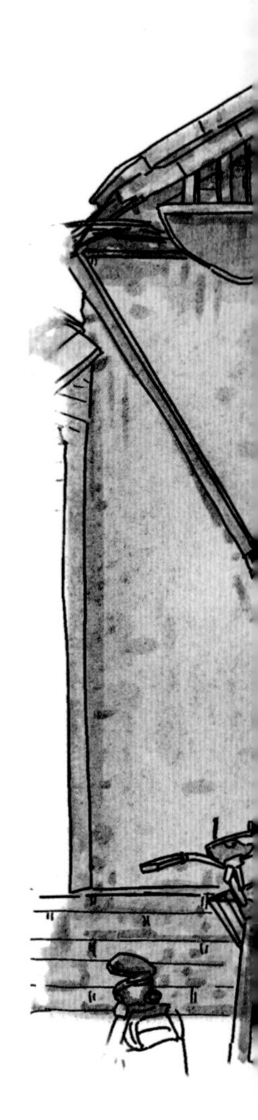

方砖厂69号
炸酱面
老北京炸酱面

06

东城区南池子大街 11 号 · 11 Nanchizi Street, Dongcheng District

去四季民福，尝一回口碑相传的“焖炉烤鸭” Taste the Duck Roasted in a Closed Oven at the Peking Chamber

四季民福的网络高口碑，是让老派北京食客“不理解”的一种火爆。说“不理解”，主要是相比北京动辄从明清传下来的几百年老字号，“四季民福”的历史实在是有点短，诞生于2008年，这在一个随便的街边小摊儿都是光绪年间传下来的北京城，四季民福别说“老字号”，连“老店”都算不上。

那您说四季民福的烤鸭靠什么火呢？没有宫廷御厨的故事，也没有皇亲国戚传说的“加持”，更没有“鹅肝酱”的中西合璧，靠的就是口碑相传——尤其是初来北京的外省朋友和吃不惯传统烤鸭的“小年轻”。我身边的外省朋友，第一次吃烤鸭，有不喜欢全聚德的，有吃不惯便宜坊的，有羞赧于大董客单价太贵的，但基本上都能接受四季民福。另外一个优点是，四季民福的价格和它的名字一样亲民：比大董和全聚德便宜不少，而且故宫店有“景观包间”，吃着鸭子能看到东华门的故宫景儿。挺好。

The Peking Chamber enjoys an astonishingly high reputation on the Internet, which confuses many older-generation gourmets in Beijing. Compared to time-honored restaurants with histories of hundreds of years, the Peking Chamber, founded in 2008, is a newcomer. Many food stalls in Beijing are much older than that. Why is the roast duck from Peking Chamber so popular? It boasts neither legends related to the Imperial Kitchen and royal family nor a fusion of Chinese and Western culinary styles. Its reputation spread by word of mouth, especially among outlanders and youngsters. When they tried Beijing roast duck for the first time, some of my friends didn't like Quanjude or Bianyifang, and others thought Dadong was too expensive, but almost all of them recognized the Peking Chamber positively. Another reason is its affordable price, much cheaper than Dadong and Quanjude. Moreover, you can see the Forbidden City through the windows of the restaurant while enjoying roast duck there.

四季民福烤鸭店
PEKING CHAMBER
四季民福

07

朝阳区麦子店街 6 号楼北侧 · North Zone, Building 6, Maizidian Street, Chaoyang District

去宝源饺子屋，宫保鸡丁也能做饺子馅儿？
Experience Dumplings Stuffed with Kung Pao Chicken at Baoyuan

我的天啊，一开始我是非常不理解宝源饺子馆的菜单的。同样是上榜米其林“必比登”推荐的北京饺子馆儿——他家的饺子馅儿如果要赶上家里传统的长辈，基本属于“乱棍打死”的级别：宫保鸡丁馅儿饺子、比萨馅儿饺子、芝士馅儿饺子、紫甘蓝锅巴馅儿饺子……您猜对了，来这儿尝一回“中国饺子”的老外特别多（它确实更适合外国朋友的味蕾——偏甜口儿改良中餐），来打卡尝鲜儿的年轻人也是乌泱乌泱排队。人生嘛，就是要勇敢地尝试。比萨馅儿饺子总得自己亲自尝一回，然后发个朋友圈是不?

您要是只能接受咸馅儿的老北京传统饺子也没关系，果断点羊肉大葱馅儿——地道，香飘两米，馋死邻桌瞎点创意馅儿“翻车”的。

I was shocked when I saw the menu of Baoyuan, a restaurant specializing in Chinese dumplings rated in the MICHELIN Guide. It offers a wide range of dumplings with untraditional stuffing: Kung Pao chicken, pizza, cheese, red cabbage... Rather than Chinese, they cater more to foreign tastes. No surprise that many foreigners frequent this restaurant. But its peculiar options also attract youngsters with an adventurous spirit. Eating dumplings stuffed with pizza is something worth sharing with your friends on WeChat Moment.

If you can only accept traditional tastes, dumplings stuffed with mutton and scallion are the top menu item, and their mouth-watering smell may arouse envy of diners sitting at the next table with exotic flavors.

屋
BAO YUAN DUMPLING RESTARANT

08

西城区六铺炕一巷 65 号 -8 · 65-8 Liupukang No.1 Lane, Xicheng District

去金生隆，天啊！北京爆肚儿还带中西结合的？
Enjoy Boiled Tripe Fusing Chinese and Western Styles at Jinshenglong

很多初来北京的朋友，可能都听过爆肚冯，对于金生隆反而有点儿陌生。

其实，金生隆作为专营爆肚的“老字号”已享誉京城上百年。清朝末年，山东厨师冯天杰来京，在东门大街摆摊卖爆肚，因其做工精细，人送外号“爆肚冯”。冯天杰的次子冯金生13岁随他学徒，16岁接替其父经营生意，并用其名后的两个字“金生”为这个店起了个正式的字号“金生隆”。

现在金生隆的掌门在日本勤工俭学过，在日料店、西餐店、咖啡馆都打过工，所以他家除了堂食的各色爆肚儿、散丹，还有需要提前预约的主厨定制菜单——比如羊肉汉堡烙馍卷儿。听起来是不是很霸气？可以去试试。

When it comes to *baodu* (boiled tripe), many first-time visitors to Beijing have heard a lot about Baodu Feng, but little about Jinshenglong.

In fact, Jinshenglong is a century-old *baodu* restaurant in Beijing. In the late Qing Dynasty, Feng Tianjie, a cook from Shandong, moved to Beijing and set up a food stall specializing in *baodu* on Dong'anmen Street. The *baodu* he cooked boasted superb culinary skills, earning him the nickname "Baodu Feng." His second son, Feng Jinsheng, started to learn cooking *baodu* at the age of 13. At 16, Feng Jinsheng succeeded to the business of his father, and officially named his *baodu* restaurant Jinshenglong.

The chef in today's Jinshenglong once studied in Japan, where he worked part-time in Japanese-style and Western-style restaurants and cafés. For this reason, in addition to a variety of *baodu*, Jinshenglong also offers an Omakase-style menu with foods fusing Chinese and foreign culinary styles such as mutton hamburger with baked bun. It sounds special, doesn't it? You may give it a try.

09

西城区平安里西大街与赵登禹路交叉口西南角
Southwest corner of the intersection of Ping'anli West Street and Zhaodengyu Road, Xicheng District

去牛街满恒记，体验一回老北京铜锅涮肉的“阵势”

Experience Beijing-Style Copper Hotpot at Manhengji on Niujie Street

满恒记属于牛街的老北京铜锅涮肉，如果您是不爱吃牛羊肉的人，很难理解“老北京铜锅涮肉”这种烹饪方式，清汤锅底不是谁家都一个味儿吗？牛羊肉不是找食材好一点儿的店都一样吗？对于挑剔的食客来说，火锅和火锅、牛羊肉和牛羊肉，里边的区别“差之毫厘，谬以千里”。小羊羔羊尾油要先下——是用来“肥锅”的。

当然，他家的热搜第一不是“点什么肉好吃”，而是“排队攻略”。别说头一次来涮肉的外省朋友，就是常来牛街的老饕，也被他家排队的“阵势”吓了一跳：现场排队根本没戏，App（应用程序）预约排号，高峰期不亚于某国民软件抢火车票。最辉煌火爆的时期，晚预约几分钟，恨不得前面多了三百个号。为什么要把满恒记放在第四章呢？带外国朋友排一回队，就是要让他们见识见识，咱中国人对一顿火锅的热情、对生活的热爱，外加上“人多力量大”的氛围感。

Manhengji is a Beijing-style copper hotpot restaurant on Niujie Street. Those who dislike beef and mutton are probably not fans of Beijing-style copper hotpot. This kind of hotpot features clear soup, and the only difference is the quality of meat. But for tricky gourmets, any tiny distinction can make a big difference. There are strict procedures to enjoy Beijing-style copper hotpot. For instance, fat lamb tail should be boiled first to make the soup “oily.”

Of course, the most asked question about Manhengji isn't what kinds of meat taste better, but how to find a seat effectively. Indeed, both tourists and locals would be shocked by the long line outside the restaurant. It is almost impossible to be seated without waiting, and you should book via its mobile app. At peak hours, it feels like booking a train ticket before the Spring Festival travel rush, with an influx of hundreds of orders within just a few minutes. Perhaps you can bring foreign friends to witness how obsessed Chinese people are with hotpot, which testifies to their zest for life.

牛街
满恒記
清真

10

西城区前门外粮食店街3号（大栅栏南）· 3 Liangshidian Street (south to Dashilanr), Qianmen, Xicheng District

去前门六必居，约半斤老北京酱菜“麻仁金丝”
Buy Pickled Kohlrabi with Sesame Seeds at Liubiju in Qianmen

看过《大明王朝1566》的朋友，一定对严嵩奉旨将“六心居”添一笔，改成“六必居”的桥段印象深刻，连电视剧里的嘉靖皇帝都记得六必居的“八宝酱菜”。故事当然是演绎的，但六必居从明代中叶以来的老字号赞誉，却是实打实的。老事年间，老北京家里做炸酱面的黄酱、喝粥的咸菜，甚至是散装酱油、糖蒜……都是六必居的。您要问，这不就是咸菜吗？还能做出花儿来？别说，六必居的酱菜技艺，还真的入了国家级非物质文化遗产代表性项目名录了。直到现在，您要是找个老北京住户多的小区，随便敲开一排单元门做个调查问卷，基本上家家户户的橱柜里，多少都有六必居。

虽说网上也能买，但来前门老店逛逛，还是有好处：一是酱菜可以散约，您可以每样来点儿——名气最大的当然是“八宝酱菜”，不过“八宝酱菜”偏甜，不爱吃甜口儿的朋友，我个人更

推荐“麻仁金丝”；至于第二大好处……老字号的实体店为了吸引年轻人，还出了糖蒜口味的“黑蒜冰激凌”！天啊，第一回听到这个消息要炸裂了。对不起，确实是为年轻人准备的？嗯，非常优秀，欢迎各位恩爱的小情侣，各自干完一盒，哈哈气，打个饱嗝儿，然后彼此谁也不要嫌弃谁。

Those who have watched the TV drama *The Ming Dynasty in 1566* probably remember a plotline involving Prime Minister Yan Song changing the name Liuxinju to Liubiju by adding a stroke according to an emperor's decree. In the TV drama, even Emperor Jiaqing of the Ming Dynasty spoke highly of the Babao pickled vegetables produced by Liubiju. Of course, the plotline is fictional, but Liubiju did originate in the mid-Ming Dynasty. In the past, Liubiju's soybean paste, soybean sauce, and pickled vegetables could be found in every Beijing household. Don't underestimate pickled vegetables. Liubiju's pickled vegetable making technique has been listed among national intangible cultural heritage of China. Even today, if you conduct a survey among older-generation Beijingers, you'll find that most still have Liubiju's pickled vegetables in their kitchens.

Although one can now buy Liubiju's pickled vegetables online, its store in Qianmen is still worth visiting for the bulk products. The most popular remains the famous Babao pickled vegetables, which taste a bit sweet. For those who don't like sugar, I recommend pickled kohlrabi with sesame seeds. To attract younger customers, Liubiju stores offer unique sugared garlic ice cream. Just imagine that a couple eat such ice cream together and then exhale garlic aroma all over each other… What a hilarious scene!

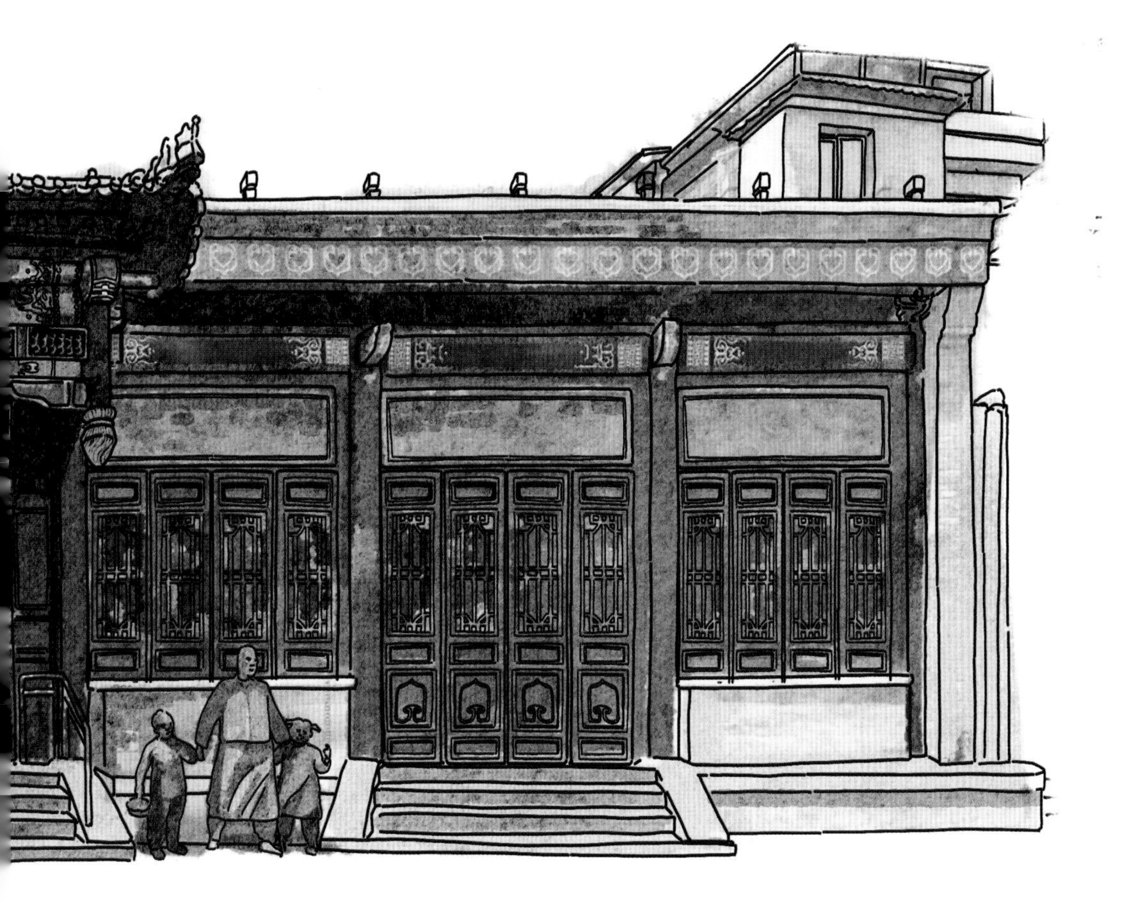

本章属于“来都来了，好歹吃一回”系列。这章的小馆儿，主要是地理位置太好了。您要是一边逛北京名物、一边想顺便吃一口附近的特色京味儿，以下都是好去处。

The eateries recommended in this chapter enjoy prime locations. For tourists who want to experience unique Beijing flavors, the following restaurants are worth visiting.

第五章
Chapter 5

来都来了

Home-Style Eateries

鸦儿李记
鸦儿李记涮
涮肉

西城区小石碑胡同烟袋斜街甲 75 号 · A75 Yandai Xiejie, Xiaoshibei Hutong, Xicheng District

去后海的鸦儿李记，一口烧饼咬下去，成仙了

Enjoy Baked Sesame Buns at Ya'er Liji on the Shore of Houhai Lake

真是"'鸦儿'别三日，当刮目相待"。印象中"鸦儿李记"就是个靠卖烧饼火起来的小馆儿，堪称后海最好吃的烧饼。没想到现在一搜"鸦儿李记"，天啊，这么多分店了？不仅烧饼有名，还成了后海最有人气的涮肉之一。至于"鸦儿李记"为什么叫这么奇怪的名字，后面的"李记"是带姓氏的字号；前面的"鸦儿"是地名——后海的"鸦儿胡同"。其实"鸦儿胡同"也是老北京念走嘴了以讹传讹，这地方属于后海北沿儿，本来是"沿儿胡同"。

我们回过来说吃。您要是都走到后海银锭桥了，那无论吃没吃饭，都推荐您顺道尝尝他家的烧饼，再打包带回来几个——太出名了。而且价格非常"美丽"：后海是啥地界儿？您还能找到一块钱一个的烧饼！因为烧饼走量快，基本排队买到的都是现出锅儿的，烧饼"一热顶三鲜"，一口下去，可能不仅成了仙，还决定晚饭就在这儿用了——为了烧饼，再加一顿涮肉。

In my memory, Ya'er Liji was a small eatery offering the best baked sesame buns in the Houhai Lake area. Today, it has opened more branches and become one of the most popular hotpot restaurants in the area. Its name is a combination of Ya'er Hutong and the last name of its founder, Mr. Li. Ya'er Hutong at the north edge of Houlai Lake was formerly called Yan'er Hutong, but in Beijing dialect, it pronounces like Ya'er Hutong. Gradually, its name changed into today's Ya'er Hutong.

If you visit the Yindian Bridge in the area, don't forget to taste and buy the famous baked sesame buns from Ya'er Liji. The sesame buns are incredibly cheap—only 1 yuan each. They sell so fast that you can basically get one just cooked. While you wait for the sesame buns, you might enjoy hotpot for lunch there.

菊儿人家

0 2

东城区小菊儿胡同 63 号 · 63 Xiaoju'er Hutong, Dongcheng District

去南锣鼓巷的“菊儿人家”，体验一回“北京最牛十碗饭之首”

Experience Beijing's Best Rice with Stewed Pork at Ju'er Renjia in Nanluoguxiang

南锣鼓巷的名气已经大到不用介绍——在这里听到的各种外语和各地的南腔北调，比北京话多。您要是路过南锣，又不想踩“网红店”，只想踏踏实实地吃顿地道的吃到胃里舒服的家常便饭，那欢迎来小菊儿胡同的“菊儿人家”，他家的卤肉饭不知道已经卖出去多少万碗，如果统计出来一定是个巨庞大的数字。开馆子的是一对老北京夫妇，卖双皮奶起家，卤肉饭一开始就是妈妈为自己女儿做的“营养家常饭”，没想到口碑爆棚，各大媒体电视台争相报道，新浪还把它列为“北京最牛十碗饭之首”——饭量大的话，可以免费续卤肉汁儿和饭。前提是不能浪费。

您要问味道多特别？它就是老北京家常的装修、家常的馆子、家常的一碗卤肉饭（卤肉饭并不起源于北京，它融合了全国各地的美食）。但是用料扎实，味道扎实，一碗卤肉饭里是妈妈的味道。

Nanluoguxiang is a well-known historical block in central Beijing. Every day, it becomes packed with tourists from around the world. In addition to native residents, visitors speaking all kinds of dialects and foreign languages will be found there on any given day. If you only want to have an ordinary dinner at an authentic Beijing-style restaurant rather than an Internet-famous one, Ju'er Renjia is an ideal choice. Since its establishment, the small restaurant has sold numerous bowls of rice with stewed pork. The restaurant was founded by an old couple in Beijing. Initially, it only sold milk custard with egg whites. Once, the mother cooked a bowl of rice with stewed pork for her daughter that made its way to the restaurant. To her surprise, it gained popularity quickly and even attracted reporters from TV stations and other media organizations. Chinese news portal Sina listed it as the top of the ten most famous rice bowls of Beijing. Those who order the dish can ask for a refill on sauce or rice if they're not going to waste it.

What is the secret behind its popularity? In fact, the restaurant is normal in interior decoration, and its offers are also ordinary in flavor. Rice with stewed pork can be found on the dining table of every household. The dish didn't originate in Beijing, but was introduced to the city as a combination of various cuisines. Featuring carefully selected ingredients, the rice with stewed pork offered by Ju'er Renjia reminds diners of a taste of mom's cooking.

03

东城区东四北大街 69 号 · 69 Dongsi North Street, Dongcheng District

去香饵胡同的胖妹面庄，尝一回米其林“必比登”推荐的辣面

Try the Spicy Noodles Rated in the MICHELIN Guide from the Pangmei Noodles Restaurant in Xiang'er Hutong

如果您看到大夏天北京接近40℃的暴晒天或者胡同里“踩水”的下雨天，还有坚持来香饵胡同等一碗面条的食客，可能纯因为“好奇心”想去打听打听：这是什么神仙面值得这么等？不就是一碗重庆小面吗？另一个好奇心也上来了：老北京也不吃这么辣的面啊？为啥这么火？

胖妹面庄是新北京的味蕾。您要是逛完孔庙和国子监，可以溜达几分钟到香饵胡同，尝尝米其林“必比登”推荐的胖妹面庄。胖妹面庄的招牌是地道重庆小面，偏辣（在老北京的味蕾体系中就算“非常辣”，但架不住年轻人喜欢）。开在香饵胡同这么多年，靠着一碗豌杂面，征服了北京南腔北调行人的胃。来，致敬新北京。

Many are surprised by the line for a bowl of Chongqing-style noodles outside a restaurant in Xiang'er Hutong in the scorching sun or the heavy rain. Is it really worth it? Why are the noodles from Pangmei so popular?

The Pangmei Noodles Restaurant is a newcomer in Beijing, but it has already been recommended by the MICHELIN Guide. After you finish a trip to the nearby Confucian Temple and Imperial College, you can walk there to taste its authentic Chongqing-style spicy noodles. Today, spicy foods are popular among young diners. Over the years, the noodle restaurant in Xiang'er Hutong has conquered the stomachs of diners from everywhere.

04

西城区新街口北大街 74 号 · 74 Xinjiekou North Street, Xicheng District

去吃一回老北京的“局气”
Experience Creative Beijing Cuisine at Juqi

“局气”原本是老北京话，称赞一个人仗义、做事有规矩，不耍赖。局气餐馆儿在北京开一家火一家，属于创意京菜，老味道新做法，并且上新频率很快——比如餐桌上来一个老北京“蜂窝煤”，别吓一跳，那是他家的黑米炒饭。再比如菜单上的“窝边草”，其实是兔子形状的甜点——老话“兔子不吃窝边草”嘛。

新街口店挨着德云社不远，您要是一定要来德云社打一回卡，那就捎带手去吃一回“局气”。如果有喜欢安静的朋友想错开“网红打卡地”，那就随便搜搜局气的分店，品控都不赖。实在是懒，去点个他家的日常外卖：一份“宫保鸡丁”的外卖，都要送一个铝饭盒，连一次性湿纸巾的包装纸上都印一句诗，冬天拿到手里，湿纸巾是温热的。“啊？北京的平价外卖都这么内卷了吗？”“不不，是因为人家‘局气’嘛”。

In Beijing dialect, *Juqi* means “upright and loyal to friends.” The famous restaurant chain named Juqi specializes in creative Beijing cuisine, and its offers are updated from time to time. Don't be surprised by black rice in the shape of honeycomb briquettes in the restaurant. When you read its menu, you may wonder what “Burrow Grass” is. It is a dessert shaped like a rabbit. The name was inspired by the proverb “Rabbits do not eat the grass around their burrows.”

The Juqi restaurant at Xinjiekou is close to the theater of Deyunshe, a famous cross-talk troupe. After enjoying a cross-talk performance in the theater, you can have dinner at Juqi. For those who dislike crowds, the Juqi branches are relatively quiet. All branches of the restaurant chain are worth visiting. You can also order takeout. The restaurant will provide an aluminum container even if you just order an ordinary dish like Kung Pao chicken. Its paper napkins are printed with poems and considerately warmed in winter. All of these efforts manifest what the restaurant chain's name suggests.

局氣
北京范兒
In Beijing
局氣 德雲社店
局氣
德雲社店
InBeijing

05

东城区大豆腐巷报房胡同小区 43 号
43 Baofang Hutong Residential Area, Dadoufu Lane, Dongcheng District

去报房胡同饺子馆，体验一回“躺平饺子”
Taste Homemade Chinese Dumplings in Baofang Hutong

这家饺子馆专治“在网红店吃出内伤的食客”。现在北京二环以里，您想找一家“网红店”遍地开花，想找一家不靠装修、餐单、晒单打8折吸引顾客，而且没有各种奇葩馅儿的家常饺子馆儿，反而成了难事。像报房胡同饺子馆这样的小店，成了如今的“重点保护对象”。

您要是去东四大街、首都剧场逛完了赶上饭点儿，可以找找报房胡同儿里的“苍蝇小馆”。无论什么时候来吃都是那个“老味道”——饺子三两起卖，只有几种馅儿：点单率最高的是猪肉茴香、猪肉大葱、猪肉韭菜。您想要的那些“网红馅儿”，统统没有！一到了吃饺子的北方节日就排大队，就是这么任性。人生味道嘛，以不变应万变，这就是老北京报房胡同的饺子哲学。

Without attractive interior decoration, menu and discount coupons, the Dumpling Restaurant in Baofang Hutong sharply contrasts many Internet-famous restaurants, making it “eccentric” among its counterparts within the Second Ring Road of Beijing. Today, small eateries like this have become “endangered” in Beijing.

After strolling down Dongsi Street or seeing a play at Capital Theater, this dumping restaurant in the nearby Baofang Hutong would be convenient. It only features a few offers, of which the most popular are dumplings stuffed with pork and fennel, dumplings stuffed with pork and scallion, and dumplings stuffed with pork and Chinese leek, all common categories. Dumplings are necessary for many traditional festivals in China, so one can often see diners queuing up outside the restaurant on such occasions. The secret to its popularity is preservation of the authentic Beijing flavor.

报房饺子馆
报房饺子馆

06

东城区雍和宫大街155号 · 155 Yonghegong Street, Dongcheng District

去雍和宫大街的新和小馆，恭喜您排上号了
Compete to Get a Seat at the Xinhe Restaurant on Yonghegong Street

有个笑话，说去新和小馆想排上号也是很容易的——建议您先步行500米，去马路对面的雍和宫拜个佛许个愿。

当然，这是笑谈。不过新和小馆真是太火了。这个门面不算太大的馆子，距离我之前工作的地方步行不足300米，但是我就没在饭点儿赶上过"不用排号，里边有位"的情况。人家生意火也是有原因的，京菜的口味地道，重要的是性价比极高，丰俭由人。乾隆白菜、三鲜锅贴儿、京酱肉丝、炸藕合都是京菜里的家常菜。蓑衣黄瓜也是招牌，吃的就是这道刀工。之前本来想装一把贤惠，跟先生说："哎呀，这20块钱一份的蓑衣黄瓜，我也能做。"上来以后见到刀工傻眼了，"算了算了，我那个刀工，可能得用20斤黄瓜（都切不出来）。咱还是下馆子吧。"

Someone joked that if you want to get a seat in the Xinhe Restaurant, you better walk 500 meters to the Lama Temple to pray to the Buddha first.

Of course, the joke reflects how popular the small restaurant is. The Xinhe Restaurant is less than 300 meters from my workplace, but I've never secured a seat here without waiting in dinnertime. Many factors fuel its popularity: It offers authentic Beijing dishes at reasonable prices. In addition to common dishes from Beijing cuisine such as Qianlong cabbage, pan-fried dumplings with shredded seafood, sautéed shredded pork in sweet bean sauce and deep-fried bamboo shoots stuffed with pork, the restaurant is also noted for sweet and sour cucumber, a dish requiring superb cutting skills. I once bragged to my husband that I could cook the dish myself. But when I saw the sweet and sour cucumber from the Xinhe Restaurant, I knew I could never cut cucumbers that well. "Uh... Sweetheart, I overestimated my capabilities, and we should dine in the restaurant." I said to my husband.

新和小馆
面筋卤香锅中
推
拉
155

07

东城区朝阳门内大街79号-2 · 79-2 Chaoyangmennei Street, Dongcheng District

去东四民芳餐厅，尝一回老北京发小儿“重口味零食”
Awaken Childhood Memories at the Dongsi Minfang Restaurant

来，再跟着念一个发音：炸灌肠（chang，轻音）。重音在“灌”，不在“肠”。灌肠不是肉菜，而是面做的。民芳是地道的北京家常菜，交情够“瓷实”的“发小儿”们才来——味道吃不腻，就像是一起长大的小伙伴儿，再聊多久都不腻。

吃炸灌肠这种需要蒜汁儿的“重口味零食”，就适合打小儿穿屁帘儿一起长大的朋友一起撮一顿。卸下你那些“妖娆”的Selina（塞利娜）胸卡、“成仙儿”的汉服或者八尺长头衔的名片，嗨，打小儿谁嘴里没有蒜味儿啊！

Zhaguanchang (literally “fried sausage”) is not actually meaty, but made of dough. The Dongsi Minfang Restaurant offers authentic Beijing cuisine, so its frequent customers are natives. For Beijingers, the restaurant is like a lifetime childhood friend.

Zhaguanchang should be eaten alongside garlic sauce. There, you should take off your work badges and leave home your beautiful *hanfu* (traditional Chinese clothing). You won’t need to show a business card to enjoy the delicacy imprinted in your childhood memories with old pals to re-experience the garlicky flavor.

四民芳餐厅
Dong si min fang Restaurant
东四民芳

08

西城区西直门内大街 132 号（近新开胡同）· 132 Xizhimennei Street (near Xinkai Hutong), Xicheng District

去天福号，称半斤乾隆三年传下来的“老北京酱肘子”

Buy Braised Pork Hock with Soy Sauce at the Century-Old Tianfuhao

天福号已经“火”到相声段子里了。北京人必须记得天福号。酱货界的“老大”——天福号酱肘子的历史传奇长得可以写本书：当年慈禧都尝过他家肘子，有了“乾隆酱汁传百年，慈禧腰牌通天下”之说。别小看酱肘子的工艺，现在天福号酱肘子的技艺和配方，属于国家级非物质文化遗产。

您要是路过西直门，可以“捎带手儿”去一趟天福号总店——百年老字号，老北京熟食。虽然现在网上也能买到真空包装的礼盒，但还是有很多人慕名来西直门总店，“来块儿酱肘子”，味儿正。

A top brand of braised pork products, Tianfuhao is embedded in the memories of Beijingers. Its legendary history is worth a standalone book. Tianfuhao was founded in 1738, the third year of the reign of Emperor Qianlong in the Qing Dynasty. Legend goes that Empress Dowager Cixi liked its braised pork hock with soy sauce very much and granted Tianfuhao a special permit allowing it to deliver the product into the Forbidden City at any time. Don't underestimate its cooking skills. Tianfuhao's traditional techniques and recipe for making braised pork hock with soy sauce have been inscribed on the list of national intangible cultural heritage of China.

If you pass Xizhimen, you can visit the century-old flagship store of Tianfuhao. Although vacuum-packed products from Tianfuhao are now available in online stores, many still prefer to buy from its headquarters near Xizhimen because the freshly cooked items taste better.

天福號
店久十万朝乃诚信为本
九州地唯品质是天

09

通州区玉桥中路 43 号 · 43 Yuqiao Middle Road, Tongzhou District

去紫光园，尝一回镇店的百年京菜“炒疙瘩”
Taste Stir-Fried Starch Knots at the Time-Honored Ziguangyuan Restaurant

如果您家里来了客人，不知道吃什么，或者想随便叫一家既“不踩雷”又有北京特色的外卖，强烈推荐老字号“紫光园”。紫光园成立于民国元年（1912），是实打实的百年老招牌，地道北京菜——品控稳定到让人忘了岁月的流逝：无论你是二十岁还是三十岁，他家的炒疙瘩一直是一个味儿。

紫光园的价格相对亲民，外卖窗口每天排大队——酱牛肉、酱牛腱、筋头巴脑……各种熟食现在都能做成真空包装礼盒。

Whether feasting guests or ordering takeout, Ziguangyuan is one of the best options. Established in 1912, Ziguangyuan is a time-honored Beijing-style restaurant with a history of more than 100 years. The quality of its offers are so stable that diners may forget the passage of time. You'll find the taste of the stir-fried starch knots from the restaurant identical to decades ago.

In addition, the prices are affordable, so there is always a long line of customers waiting for seats. Its offers include braised beef with soy sauce, braised beef shanks with soy sauce, stewed beef tendons and many others. All can be vacuum-packed.

紫光园

10

朝阳区朝阳北路水碓子小区 3 号楼下南平房
South Bungalow, Building 3, Shuiduizi Residential Area, Chaoyang North Road, Chaoyang District

去平安小馆儿，尝一回北京老百姓口味的地道锅贴儿
Taste Authentic Beijing-Style Pan-Fried Dumplings at Pingan Xiaoguan

可能要让旅游打卡一族的朋友略感失望，平安锅贴的老店并不属于任何旅游景区范畴——这是一家藏在居民楼里的京菜小馆儿，离呼家楼、金台路都不到一站地。如果恰好去附近办事儿，倒是可以沿着导航路线到水碓子小区，找到平安锅贴的老店。锅贴都是家常馅儿：猪肉三鲜、羊肉西葫芦、韭菜鸡蛋等。如果爱吃肉的朋友可以来个酱肉拼盘——家常口儿略咸，巨豪横。

啥？您不知道什么是“锅贴儿”？嗯……类似于北方版的“生煎”，或者拿英文直译可能您

更容易理解：Pan-fried dumplings（一种用平底锅煎出来的、类似于饺子的吃食）。有画面感了吧？宇宙的尽头是中国菜，中国菜的尽头是各种饺子。

Pingan Xiaoguan may let down those seeking an instagrammable restaurant. It is merely a nondescript Beijing-style eatery hidden in a residential building close to Hujialou and Jintailu bus stops. If you happen to pass it, you can visit to taste its famous pan-fried dumplings. Common categories include dumplings stuffed with pork and seafood, dumplings stuffed with mutton and marrow squash, and dumplings stuffed with Chinese leek and egg. For those who like meat, they can order assorted meat braised with soy sauce, which tastes a bit salty but yummy.

As the name suggests, pan-fried dumplings refer to dumplings cooked in a frying pan. Someone joked that the endpoint of the culinary universe is Chinese food, and the endpoint of Chinese food is the various kinds of dumplings.

东城区北剪子巷 21 号悠惠万家超市内（北剪子巷与桃条胡同交叉路口）· Youhui Wanjia Supermarket (at the intersection of Beijianzi Lane and Taotiao Hutong), 21 Beijianzi Lane, Dongcheng District

去大华煎饼，尝一回老北京胡同里的“薄脆天花板”

Taste the Best Pancakes at Dahua, Hidden in a Beijing Lane

人生有很多不能理解的事——比如北剪子巷的大华煎饼，一个连堂食都没有的小门脸儿，居然排队排到很多人“空手而归”：得上午去，下午再排队就卖没了。煎饼摊儿位置难找到需要用“北剪子巷与桃条胡同交叉路口”来概述。您要是路过交道口，一定得骑个单车钻个胡同儿，来尝一回他家的招牌煎饼。

不过要提醒一点的是，他家的煎饼属于新派口味：煎饼里可以加芝士，是甜口儿。这要是搁老北京的胃里，够“枪毙”五分钟的——这是“传统咸鲜党”的叛徒。不过，架不住南来北往的年轻人喜欢。另外不得不说，薄脆炸得太好了，天花板级。

Many can’t understand why the pancakes offered by a food stall named Dahua are so popular. Countless customers queue up to wait for a pancake every day. Even so, some leave empty-handed. To get one, you must join the line earlier because the pancakes sell out by the afternoon. If you pass the intersection of Beijianzi Lane and Taotiao Hutong by bicycle, visit this food stall hidden in a narrow alley to taste its pancakes.

It is worth noting that its new-style pancakes contain cheese and taste a bit sweet, which is considered “unacceptable” by loyal fans of traditional salty pancakes. However, this new flavor is popular among young people from around the country. Also, Dahua’s crispy fritters are absolutely delicious and dubbed the best in Beijing.

大华煎饼
DAHUA PANCAKE
大華煎餅

12

门头沟区大峪街道新桥大街78-1号 · 78-1 Xinqiao Street, Dayu Sub-district, Mentougou District

去新桥炸鸡店，点一份儿时味道的“老式炸鸡”

Experience Traditional Fried Chicken from Childhood at the Xinqiao Fried Chicken Restaurant

如果您路过门头沟区，大老远来一趟，最好开一下地图导航，去新桥炸鸡店尝一回“老式炸鸡”再走，才算是不辜负来此一游。新桥炸鸡店属于老式炸鸡——就是80后、90后儿时记忆中肯德基还算是“考好了才能奖励的奢侈品”时代的炸鸡平替。老式炸鸡的孜然、辣椒粉要吃的时候单独撒，味道咸口儿，刚出炉趁热的时候外皮酥香解馋，里边的鸡肉嫩滑入味。要提醒的有两点：一是老店的“老味道”往往偏咸，过去的人都觉得有油盐味儿才香，和现代低盐低钠的新派饮食观点不一样；二是总店排队的人多，您要是开车来，一定要把车找地方停好了再排队。不然回来一看，炸鸡五十，贴条罚款二百。伴着心疼的眼泪吃着炸鸡，分不清是眼泪咸还是料包咸。

If you travel to Mentougou District in suburban Beijing, you should drop in on the Xinqiao Fried Chicken Restaurant to taste its famous old-style fried chicken. It is absolutely worth it. Back in the 1980s and 1990s when KFC remained a "luxury" reward for kids with outstanding performance in school exams, the Xinqiao Fried Chicken emerged as an affordable alternative. The old-style fried chicken tastes a bit salty, and you can add condiments such as cumin seeds or pepper powder. Freshly fried chicken is crispy outside and tender inside. Two things to know: First, old-style cuisines are usually salty, in opposition to modern diets that emphasize less intake of salt and sodium. Second, the flagship branch of the Xinqiao Fried Chicken is always crowded with diners, and customers usually need to wait in line for a seat. Park in an authorized lot or you might get a parking ticket if you leave your car nearby. The fine will be much more than what you spend on fried chicken.

新桥炸鸡店
新桥炸鸡店
新桥大街直营店

13

西城区大秤钩胡同甲 9 号 · A9 Dachenggou Hutong, Xicheng District

去大秤钩胡同，尝一回地道的“老北京馅饼粥”
Taste Authentic Beijing-Style Pie and Porridge in Dachenggou Hutong

现在的潮人开店，八字还没一撇，都得先准备九个“店名”外加十页幻灯片演示，这才叫“以店名反哺目标生态，赋能整体业务，击穿用户心智”。老店才不理会这一套：比如西单“老北京馅饼粥”——人家的店名就是“没店名”：只有一行大字“老北京馅饼粥”。

因为太没记忆点，搜店名都不好搜，附近居民干脆简称“西单大馅饼”。在这里，馅儿就是传统馅儿，点单率最高的猪肉大葱馅儿和牛肉馅儿，吃素的朋友可以选韭菜鸡蛋馅儿。皮薄馅儿大，用料扎实，出锅的馅饼油香四溢，胃小的女生基本一个就饱了。您要说是人间至味，小店肯定比不上豪宴山珍，但那股透出来的家常亲切感和店主地道的北京话，让人觉得，这大概是环游世界一圈儿回来，阅尽千帆，大道至简，发现家门口儿一口家常馅儿饼才是“至味”的美食哲学。

Before opening a modern restaurant, the proprietor might prepare a lengthy list of possible names and a business presentation to explain various Internet slangs such as "regurgitation-feeding the objective ecology, empowering the overall business, and targeting the hearts of users directly with an eye-catching and soul-stirring name." However, time-honored restaurants are the opposite, and some don't even have a formal name but instead just list the primary offers. For example, "Beijing-style Pie and Porridge."

The nondescript names makes this restaurant hard to search for on the Internet. Locals often call it "Xidan Big Pie." The pies it offers feature traditional fillings, of which the most popular are pies stuffed with pork and scallion and pies stuffed with beef. Vegetarians can choose pies stuffed with Chinese leek and egg. The pies boast thin crusts and adequate filling, with an overflowing aroma of cooking oil. A girl with a small stomach might be full after eating only one such pies. Of course, this small eatery cannot compare with luxury restaurants in either flavor or decoration, but the Beijing dialect of its owner gives diners a sense of intimacy. After tasting delicacies from around the world, they eventually realize that the simple pies from home are what they crave.

老北京馅饼粥

北京话的“瓷”，就是形容极好的朋友，用现在的流行话就是“老铁”。

生活需要仪式感，吃饭也不例外。纪念日、特殊的日子、要好的朋友……总有一些人值得我们在“吃饱”之外的餐厅环境、菜品意境、细节服务、文化高度等细节中用心思。这一章“意境菜”，基本不是用来吃的，是用来在特别的日子，和值得的人一起“赏”的。

In Beijing dialect, *ci* means “good buddies” like the recent buzzword *Laotie*.

Life requires a sense of ritual, and dining is no exception. Anniversaries, special days, and gatherings with friends all demand food. Some occasions are always spent with loved ones in carefully selected restaurants not only with a fancy environment and delicious food, but also an artistic touch and pleasant services. At such places, diners come for more than just food and also tend to enjoy leisure time with special people.

第六章
Chapter 6

瓷，咱吃的不是菜，是「意境」

More than Just Food

01

西城区阜成门内大街 162 号 · 162 Fuchengmennei Street, Xicheng District

去和芳苑，在四进四合院的榜眼府第，喝一回下午茶

Enjoy a Traditional yet Fancy Afternoon Tea in the Courtyard of Hefangyuan

来这儿吃饭喝茶的，都不是俗人。俗人吃饭讲究实惠，吃到肚子里才算。和芳苑嘛，您先看这院子就知道大概的品位了：二环以里，还能找到完整四进四合院的私房菜府邸，真是太难得了。和芳苑之前是同治十三年（1874）榜眼谭宗浚的府邸——现在还保留着四进四合院的亭台楼阁、雕梁画栋和曲径通幽。民国时期北京城“食界无口不夸谭”，指的就是谭家菜。而和芳苑，算是谭家菜的发源地——谭府佛跳墙、谭府黄焖翅都是道道有讲儿的招牌。

除了正餐，您还可以用宫廷点心配谭府下午茶，园子一步一景，不到谭府，怎知春色如许？当然了，人均消费也不便宜，正餐人均1300元起步，下午茶人均800元一位。人生重要的朋友不多，如果遇上了，那值得来谭府，赏一回春光。

Hefangyuan, a fancy restaurant hidden in a traditional courtyard, is for those with elegant taste and little concern for cost-value. Today, it is rare to find a restaurant offering private home cuisine within a traditional *siheyuan* (quadrangle residence) within the Second Ring Road in Beijing. Hefangyuan was formerly the mansion of Tan Zongjun, who passed the imperial examination as the *bangyan* (second place) in 1874, the 13th year of the reign of Emperor Tongzhi (1862-1874) of the Qing Dynasty. Within four courtyards, the restaurant still maintains pavilions, towers, and other traditional buildings with painted beams and decorative carvings as well as zigzagging paths. During the Republic of China period, Tan Family Cuisine earned a high reputation in the culinary community in Beijing. Hefangyuan is the birthplace of Tan Family Cuisine, and its signature dishes include Tan's Fotiaoqiang (steamed abalone with shark's fin and fish maw in broth) and Tan's Braised Shark's Fin with Brown Sauce.

In addition to main meals, visitors can also sip afternoon tea while enjoying spring views in the former Tan's mansion. Of course, the prices are high. A main meal costs at least 1,300 yuan per person, and afternoon tea is about 800 yuan per person. True friends come around once in a lifetime, and those are the kind of people worth taking to this place to enjoy leisure time together.

和芳苑

02

东城区东四十一条25号 · 25 Dongsi No.11 Lane, Dongcheng District

去曲廊院，感受一回四合院里的“中西交融艺术菜”
Experience Artistic Dishes Fusing Chinese and Western Styles at Qulangyuan

曲廊院也是在寸土寸金的地界儿——东四十一条胡同25号——能“趁”一个明清四合院的餐馆。这儿的每道菜，都是一半餐一半艺术：有的摆盘像水墨画，有的摆盘像浮世绘，有的摆盘像文人小品。菜单也是中西结合、新旧交融：既有中式的烤羊羔排，也有洋范儿的鹅肝松露饭。

园子的设计改造也有妙趣，屡获建筑设计奖项。

除了吃饭，在这儿还能看艺术展：水墨、版画、摄影、陶瓷……当然了，唯一需要提醒的是，这里的消费也不算便宜，人均1500元左右。有“点餐恐惧症”也无妨，可以直接网上团券点套餐，人家都给您配好了。春日丽，夏蝉鸣，人生总有一些与值得的人一起共度的纪念日，在曲廊院用一次小宴。

Located at 25 Dongsi No. 11 Lane in central Beijing, Qulangyuan is another restaurant hidden in a traditional courtyard dating back to the Ming and Qing dynasties. Here, each dish is like a work of art. The plating itself is artistic: some look like an ink-and-wash painting, some like a Ukiyo-e print, and others like a calligraphic work. Its offers blend Chinese and Western cuisines. Not only Chinese-style grilled lamb chops but also Western-style rice with foie gras and truffle are on the menu.

The design and renovation of its courtyard has won several architectural awards.

In addition to catering, you can also enjoy exhibitions of ink paintings, printmaking, photography, ceramics, and other arts. Of course, it is worth noting that the food is expensive here, with an average per person price of about 1,500 yuan. For those with a "phobia of ordering," they can order online. On a sunny spring day or a cicada-singing summer day, it's worth enjoying delicacies with loved ones at Qulangyuan.

貳拾伍號
曲廊院

03

东城区景山东街三眼井胡同丙68号（景山花园酒店内）
Jingshan Garden Hotel, C68 Sanyanjing Hutong, Jingshan East Street, Dongcheng District

去望春食阁，景山下尝一回“宫墙四重奏”下午茶

Experience the "Royal Palace Quartet" at Wangchun Shige at the Foot of Prospect Hill

如果有人求问：“想带女朋友去吃一顿能体现出四九城‘贵气’的下午茶，最好能感觉紫禁城就是自己家的，但是预算只有人均100块，请问有推荐吗?”

嗯……脑子里想了想，还真有一家符合要求的：景山附近的望春食阁。望春食阁的地界儿太好了——景山东街的三眼井胡同，二楼的露台能看到景山的万春亭。在这儿吃顿午饭，感觉景山就是“自己家”的。菜品属于融合菜，有传统京菜，也有南方菜。不要被“宫墙四重奏”的菜名儿搞晕了——四种随机拼的小菜，店家会根据你点的菜来帮你配，不会重样。咋样？找个天朗气清的日子，景山就酒，越喝越有。不过这种景山就在“眼么前”地段儿的食阁，最好去之前打电话问下老板当天是不是正常营业。

A friend once approached me for a recommendation. "I want to take my girlfriend for afternoon tea in central Beijing at a place where we might feel like we own the Forbidden City, but I can only afford 100 yuan per person," he said. "Do you know of anywhere like this?"

I thought for a while and finally recommended Wangchun Shige, a restaurant located in Sanyanjing Hutong, Jingshan East Street, close to Jingshan Park north of the Forbidden City. From the balcony on the second floor of the restaurant, one can see the Wanchun Pavilion on Prospect Hill in the park. While eating there, one might feel as though they owned the Forbidden City. The restaurant offers mixed flavors including both Beijing cuisine and those from southern China. One may wonder about the "Royal Palace Quartet" on its menu. In fact, it is a dish randomly comprising four kinds of vegetables. Have lunch with several friends in the restaurant while enjoying the scenic view of nearby Jingshan Park on a sunny day. Do call the restaurant in advance to make sure there are available seats.

04

朝阳区北辰路北土城地铁口 A 口西侧
West to Exit A, Beitucheng Subway Station, Beichen Road, Chaoyang District

路过奥体中心，尝一次梦回《东京梦华录》里的“宋宴”

Experience Song Dynasty Lifestyles Described in *The Dreamlike Glory of the East Capital* at the Song Banquet

先不用紧张，来“宋宴”吃一顿北宋的豫菜，不用考河南话。

想不到吧？有人真把宋代饮食文化“复活”了——北宋的豫菜和南宋的淮扬菜创意融合。来这儿吃饭只盯着“口味”就有点儿浪费了：餐食盛器、字画盆景……处处都做足了宋代的风雅之美：焚香、点茶、挂画、舞蹈，只缺一个机位和导演，感觉自己就能到《东京梦华录》现场了。

瞧瞧人家那盘子！当然，菜吃不完可以打包，盘子不能带走。眼睛耳朵也不能闲着：还有蒙着面纱的“宋代女子”歌舞表演——然后上了一碗地道的豫菜胡辣汤。人均消费在500块左右，穿着汉服来体验一把?

Don't worry: you don't need to speak Henan dialect to taste Henan cuisine from the Northern Song Dynasty (960-1127) in the restaurant named Song Banquet.

No one expected a restaurant to revive Song Dynasty food culture by creatively fusing Henan cuisine of the Northern Song Dynasty (960-1279) with Huaiyang cuisine of the Southern Song Dynasty (1127-1279). It would be a waste to eat at this place without understanding everything there. The tableware and ornaments such as calligraphy, paintings, and potted plants all demonstrate the elegant style of the Song Dynasty, so do incense burners, tea ceremonies, and dance performances. Every element makes diners feel like they are living in scenes described in *The Dreamlike Glory of the East Capital*, an ancient book on the lifestyles in the capital of the Northern Song Dynasty.

Even the plates are exquisite and elegant. Of course, you're not allowed to take them home. The song and dance performed by young ladies wearing veils and Song Dynasty costumes present a feast to the eyes and ears. Then, you can enjoy a bowl of authentic Henan-style pepper soup. Prepare to spend about 500 yuan per person. Why not wear traditional clothing to experience dining at the Song Banquet?

宋宴

05

东城区前门大街 50 号 · 50 Qianmen Street, Dongcheng District

去“宫宴”，穿着汉服体验一回《礼记》里的燕享食礼

Experience the Dining Ritual Recorded in *The Book of Rites* at the Palace Banquet

如果您觉得上文的“宋宴”没有过足瘾，可以来前门的“宫宴”搞一个加强版——穿着汉服体验一回《礼记》里的燕享食礼（“燕享”同“宴飨”，“燕享食礼”始自周代）。整个“宫宴”从菜品到餐具到氛围环境到表演服务，更像是一场沉浸式视觉盛宴，随便拿个手机都能拍出“大片儿”感。

来“宫宴”必须提前预约，人均消费在500元～600元之间，除了歌舞节目，还提供汉服和化妆服务（需单付费）。餐单也是源于典籍《礼记》做足了功课，菜品丰俭由人：宫廷冷盘、佛跳墙、鸽子豆腐汤……特别适合带外国朋友来体验一把：从吃饭到表演到中国文化体验，一顿饭一体式到位，省去了“尴尬癌”找话题和翻译器的时间，顺便惊掉他的下巴。

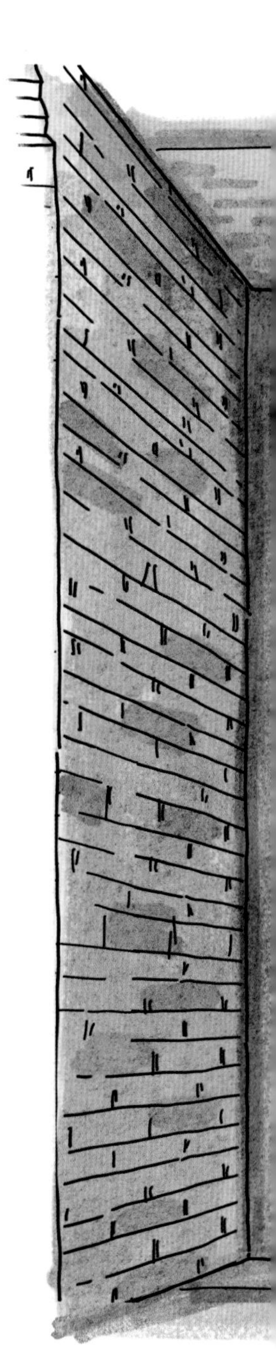

If the Song Banquet is too “contemporary” for you, visit the Palace Banquet at Qianmen, where you can experience the dining ritual recorded in *The Book of Rites* dating back to the Zhou Dynasty (1046-256 B.C.). From food and tableware to interior decoration and art performances, the restaurant presents an immersive visual feast. One can take an instagrammable selfie no matter how old your smartphone is.

One must make a reservation to dine at the Palace Banquet, where the cost is about 500 to 600 yuan per person. In addition to song and dance performances, the restaurant also offers traditional clothing rentals and makeup services. Its menu is also inspired by *The Book of Rites*. The restaurant offers a wide range of dishes at different prices including Royal Cold Dish, Fotiaoqiang, and pigeon soup with tofu. The place is particularly ideal for foreigners to experience Chinese culture while enjoying Chinese food. There, they will become immersed in fascinating delicacies and performances without the need for an interpreter.

宴宫
ALACE

06

海淀区苏州街 29 号 · 29 Suzhou Street, Haidian District

去白家大院，尝一回旧王府的“格格宫廷菜”
Taste Royal Cuisine at Bai's Compound, a Former Prince's Mansion

白家大院，听着一个蛮低调的名字，其实地界儿也是寸土寸金：之前是旧王府的园子。如果您来北京，不仅是为了“吃”，还希望“体验一回宫廷菜氛围感”，那可以来白家大院——服务员穿得都跟格格似的，见了客人都是“您吉祥”，感觉自己离贝勒爷就差一条辫子的距离。

来白家大院如果只是“低头吃饭”就亏了：园子这么大，总要走一走逛逛风景，体验当一回格格“用膳”的待遇。客人基本以商务宴请和外国朋友居多。地道北京菜——宫廷小吃、宫爆虾球、宫廷烤鸭……

Bai's Compound, which sounds nondescript, was formerly a prince's mansion. For tourists visiting Beijing, it is an ideal place to experience royal cuisine. The waitresses dress like Manchu princesses and greet customers in a traditional way, making them feel like Qing royalty.

In fact, Bai's Compound is more than just a place to eat: its spacious courtyard is ideal for taking a walk. The majority of its customers are business people and foreigners. The restaurant offers authentic Beijing cuisine including Royal Snacks, Kung Pao shrimp balls, and Royal Roast Duck.

御仙都
傳統活在當下

海淀区西四环北路117号 · 117 West Fourth Ring North Road, Haidian District

去御仙都皇家菜博物馆，尝尝“皇家菜”的非遗传承

Enjoy Cultural Heritage at Yuxiandu Imperial Cuisine Museum

现代人生活条件好了，吃饭不仅仅是“舌尖儿”的事情，还要文化、要氛围、要环境、要细节、要服务，要综合体验……前面推荐的几家馆子，包括御仙都皇家菜博物馆，严格来说都属于“七分服务三分吃”的新派沉浸体验餐厅。

御仙都皇家菜博物馆在西四环北路，不仅能吃饭还能逛博物馆——他家的“御膳制作工艺”在2012年列入非物质文化遗产名录。装修和环境满分，金碧辉煌可以直接拍电影了——进门就可以梦回古代宫廷。如果您有需求，还可以体验一回“换装服务”。要不去尝尝？

With the rise in living standards, dining became more than about eliminating hunger. It is now an integrated experience involving culture, environment, details, and service. Just like the restaurants in this chapter, Yuxiandu Imperial Cuisine Museum is a new-style immersive restaurant that prioritizes service over food.

Located on West Fourth Ring North Road, the museum gives visitors an experience with not only delicious food but also intangible cultural heritage. In 2012, its imperial cuisine making technique was listed among the intangible cultural heritage of Haidian District. Moreover, it boasts gorgeous interior decoration, making it look like an ancient palace in a movie. It also provides traditional costume rental service.

08

通州区张台路金福湿地公园内 · Jinfu Wetland Park, Zhangtai Road, Tongzhou District

去副中心的湿地公园，走一趟"能吃鱼的博物馆"
Eat Fish in a Museum-like Restaurant Hidden in a Wetland Park

如果您周末来副中心自驾游，带着一家老小逛金福湿地公园——看看大鱼，赏赏芦苇，亲近一下大自然的水，中午要是逛累了，可以去"能吃鱼的博物馆"吃顿饭。整个餐馆是红砖水泥艺术外观，乍一看真的像是湿地公园里的博物馆——一步一景的艺术装置，明窗净几的用餐区，大落地窗正对着公园湿地，可赏湿地公园秋日的一地金黄。清蒸鲟鱼是他家的招牌，鲜嫩肥美。

要提醒的是，他家是湘菜打底，不能吃辣（或者小朋友不能吃重口的）一定要提前跟服务员反复确认，千万不要逞英雄：湘菜厨子的"微辣"相当于北京菜的"忒辣"。

If you visit Jinfu Wetland Park with your family to engage in close contact with nature on a weekend, you can eat fish for lunch in a restaurant there that feels like a museum. Featuring red brick facades, the restaurant appears like an art museum. In addition to artistic installations, it also boasts a spacious dining space with floor-to-ceiling windows through which diners can enjoy scenery of the wetland park, especially in autumn. Its signature dish, steamed sturgeon, tastes fresh and yummy.

It is noteworthy that the restaurant mainly offers Hunan cuisine, which is very spicy. For those who cannot afford spicy food or those with children, remember to tell the waiter your needs when ordering dishes.

能吃的博物馆

09

海淀区花园路街道牡丹园西里 16 号楼 1-2 层（牡丹园地铁站 F 口步行 200 米）
F/1-2, Building 16 (200 meters from Exit F of Mudanyuan Subway Station), Mudanyuan Xili, Haidian District

去北平楼，重回晚清民国风的“北平民间菜”
Experience Folk Cuisine from the Late Qing Dynasty and Early Republic of China Period at the Beipinglou Restaurant

北平楼的装修风格，有点儿像四世同堂和南京大牌档——环境都是老北京的老物件儿再现，像是“穿越”回了晚清民国时期：永利达钟表行、方华美照相馆、“鸿运景福”的老牌匾、京剧的戏服行头展……等餐的时候还可以逛逛，相当于重回了老胡同——人家的服务宗旨，就是要老胡同人吃出儿时的记忆。

虽然服务员都穿着“宫里”的衣服来上菜，菜却是地道的胡同儿“北京民间菜”：除了“格格爱吃的酥肉”“王朝烤鸭”，偶尔来盘儿“鸡汁儿豌豆尖儿”、糖饼，也挺对胃。要提醒您的是，他家生意太火了，去之前最好电话问问需要等位的时间。

The interior decorations of the Beipinglou Restaurant such as signboards from Yonglida Clock Shop and Fanghuamei Studio, a plaque carrying Chinese characters *Hongyun Jingfu*, and Peking Opera costumes and props bring diners back to the late Qing Dynasty and the early Republic of China period. This style is similar to that of *Sishi Tongtang* (four generations under the same roof) and *Nanjing Dapaidang* (sidewalk snack booth in Nanjing). Such a design awakens Beijing *hutong* dwellers' childhood memories, which is just what the restaurant wants.

Although waiters and waitresses dress in palace costumes, the restaurant offers mostly authentic folk cuisine from Beijing. In addition to Princess's Crisp Pork and Imperial Roast Duck, diners can also enjoy snacks such as snow pea shoots in chicken soup and sugared sesame cakes. The restaurant is extremely popular, so remember to make a reservation in advance.

北平樓

10

东城区王府井大街王府井百货负二层
F/B2, Wangfujing Department Store, Wangfujing Street, Dongcheng District

去和平果局老北京风物馆，穿越 40 年前的“老北京胡同儿”吃食

Revive Old-Time Beijing Lifestyles at Heping Guoju

如果您带着孩子来，不仅想买老北京点心，还想找个地方玩一会儿，推荐来王府井的和平果局。这里变成了老北京风物馆——进来傻眼了，仿佛回到了老北京胡同里的童年。老式的洗脸盆、夜壶……尤其这洗脸盆还是花底儿，夜壶还是搪瓷的，侧面搭毛巾的架子，胡同儿里昏昏的灯，水泥柱的电线杆，胡同里“居民委员会”白底黑字的牌子……一个个带着孩子的大人，在这里回到了自己的童年。

临走到果局的“糕点铺”，这里各种的老北京传统点心：京八件、桃酥、手工糖果子、铁桶饼干……哎，突然鼻子一酸。要提醒的是，如果赶上暑期高峰，想逛“胡同文化节”，参观是有基础门票的。比较介意的朋友可以提前打电话问一句，咱们大老远来一趟，一定得图个开心。

If you want to buy a traditional pastry of Beijing during leisure time with your kid, Heping Guoju (literally "Pastry Shop") at Wangfujing is an ideal option. Decorated wash basins, enamel chamber pots, towel stands, old-style street lights, cement utility poles, and white signboards for community committees with black inscriptions all awaken Beijingers' childhood memories of *hutong* life.

At Heping Guoju, customers can find all sorts of traditional Beijing pastries including *jingbajian* (eight-set Beijing-style pastry), nut biscuits, handmade candies, and biscuits contained in an iron barrel, all capable of inspiring nostalgia. It is noteworthy that admission is required for the Hutong Culture Festival held at Heping Guoju each summer. Before each visit, remember to call in advance to see whether a ticket is needed.

北 京
BEI JING
和平東局 ←→ 未来

人间烟火气，怎么能少了爱吃肉的人。这一章都是地道的肉菜，恨不得肉炖肉的过瘾系列，推荐给无肉不欢的朋友。

Eating meat is an indicator of an affluent life. All restaurants mentioned in this chapter are noted for their meaty dishes, and meat gourmets regard them as paradises.

第七章
Chapter 7

肉食动物

Paradises for Meat Gourmets

01

东城区雍和宫大街 185 号（近簋街西口）
185 Yonghegong Street (near west entrance of Guijie Street), Dongcheng District

去烤肉宛，“文吃”师傅的手艺同样“豪气”
Experience “Literary Dining” at Kaorou Wan

很多人来烤肉宛，都是看了非遗纪录片慕名来打卡的。北京烤肉“南宛北季”，光知道烤肉季岂不是俗气了？会吃的人还得来吃个烤肉宛。怎么区分两家烤肉呢？烤肉季祖上是卖烤羊肉的，而烤肉宛从康熙年间起，祖上是卖酱牛肉起家的，属于北京最资深的老字号之一——头回来的食客，推荐尝尝他们家的牛肉。

在烤肉宛，文吃武吃皆可。不想出一身汗费力气的话，体验一回“文吃”师傅的手法也是艺术享受：肉片腌得恰到好处，切得薄如蝉翼。老炙子、松木火，少烟有香气。要提醒的是，如果您去的是簋街老店，老城区建筑改造规矩多，他家上二楼还是需要非常传统的方式——爬楼梯。如果是带着老人或小孩儿来，特别介意“没有电梯”的，可以提前问一问服务员。青壮年就无所谓了：多走两步路，吃得香啊。

Many customers visit Kaorou Wan to experience its barbecue, which has been listed as an intangible cultural heritage item. Beijing has two time-honored barbecue restaurants: Kaorou Wan in the south and Kaorou Ji in the north. What is the difference between the two restaurants? Kaorou Ji started with grilled mutton, while Kaorou Wan started with beef stewed in brown sauce. Founded in the reign of Emperor Kangxi (1662-1722) of the Qing Dynasty, Kaorou Wan is one of the oldest existing restaurants in Beijing. First-time visitors are suggested tasting its grilled beef.

Both “literary dining” and “martial dining” are available in Kaorou Wan. If you don’t want to grill the meat yourself, you can experience “literary dining”—namely, asking the cook to grill the meat. The cook boasts state-of-the-art cutting skills, and the meat he slices is as thin as a cicada’s wings. Traditional grills are put on burning pine firewood, which produces less smoke while sending forth an aromatic smell. One thing you need to know is that the restaurant in Guijie Street has no elevator, and customers have to climb a staircase to the second floor, which is a bit inconvenient for elders or toddlers with a difficulty in walking. Of course, young adults won’t care. A bit walk can whet the appetite.

烤肉宛
烤肉宛
京城清真第一烤
放心就餐到华天

02

西城区阡儿胡同 41 号 · 41 Qian'er Hutong, Xicheng District

去烤肉刘，找找汪曾祺散文里的“主人公”
Rediscover the Protagonists in Wang Zengqi's Prose at Kaorou Liu

如果您小时候看过汪曾祺的《捡烂纸的老头》——全篇都是用北京话写的，讲的就是虎坊桥烤肉刘馆子里的事儿。烤肉刘也是北京烤肉的老字号——1900年创立，他家的烤肉腌制配法是非遗传承，不蘸料都好吃。烤肥牛、烤羊肉基本是桌桌必点的招牌，也有重口儿的朋友要尝尝炸窝头配臭豆腐：嗯，越吃越香，来的人必须每人吃一筷子，这样大家各不嫌弃——尤其是吃美了特别想打个舒服的饱嗝儿的时候。

如果下雪天儿去就更好了，体验一回《明宫史・饮食好尚》中“凡遇雪，则暖室赏梅，吃炙羊肉”的雅气。喜欢体验胡同儿文化的，可以去汪曾祺笔下的虎坊桥老店——写到这儿突然想起来统一叮嘱一句，胡同儿里的老店大部分都得走几步去公共卫生间。这也是胡同文化的一部分不是？卸下来滤镜的北京胡同儿，才是真胡同。

If you read Wang Zengqi's "An Old Man Who Collects Waste Paper for a Living," which features Beijing dialect, you will know the story it tells happened at Kaorou Liu in the Hufangqiao area. Established in 1900, Kaorou Liu is a time-honored barbecue restaurant in Beijing. Its barbecue seasoning recipe has been listed as an intangible cultural heritage item. Grilled meat even doesn't need to add any extra condiments. Almost every customer will order the restaurant's signature offers—grilled beef and grilled mutton. Those who like heavily flavored food can try fried corn bread with stinky tofu, which smells bad but tastes good.

It is better if you dine here on a snowy day, just as described in *History of the Ming Palace*: "Eat grilled meat in a warm chamber while enjoying plum blossoms outside when snow falls." How pleasant it is! Those who like to experience *hutong* life are suggested visiting the old restaurant near Hufangqiao mentioned in Wang Zengqi's work. I have to remind you that there is no restroom in the restaurant, and you will have to use public toilet in the nearby *hutong*." In fact, this is a part of authentic *hutong* life in Beijing.

烤肉刘
烤肉劉
垂涎百尺烤肉劉
難解珍饈和美味

03

海淀区甘家口北街甘家口 21 号楼 1 层底商 · F/1, Building 21, Ganjiakou North Street, Haidian District

去柴氏风味斋，吃一碗米其林“必比登”榜上的“小碗肉”

Eat Small Beef Bowl Rated in the MICHELIN Guide at Chai's Fengweizhai

做小碗肉的最高境界是什么？就是得让不爱吃肉的人都能给好评：说明牛肉处理得好，一点儿杂腥味没有。其实柴氏最早的招牌就是牛肉面——“灵魂”在汤和肉。他家的小碗肉、酱牛肉都属于“必点”，尤其是小碗肉，牛肉的料和火候都足够，浮油已经全炖出去了，不爱吃肉或者不吃一点儿肥肉的人都能狠下几筷子。

如果您觉得肉菜点多了有点儿腻，可以搭配一个“焯菜心”。有一说一，有时候最怕在北京馆子招待广东朋友，常常被“鄙视”——北京的“焯菜心”不能用粤菜的“白灼”去严格要求，就是开水焯一下，属于家家户户做面条的配菜，要的就是一个去腻爽口的清淡口儿。对，没滋味就对了。日子至高境界，莫嫌没滋味。

What is the best Small Beef Bowl? It can even earn praise from those who dislike meat. That means the beef is carefully processed, without any unpleasant flavor. The original signature offer of Chai's Fengweizhai is beef noodles, and the soul of this dish lies in the soup and high-quality beef. Other must-try dishes include Small Beef Bowl and braised beef in brown sauce. Small Beef Bowl, in particular, boasts carefully selected and cooked beef, which is palatable but not greasy. Even those who are not fans of meat will like it.

If you don't like too much meat, you can order boiled green vegetables. To be honest, sometimes I'm afraid of treating friends from Guangdong in a Beijing restaurant because they will complain that the Beijing-style boiled green vegetables is a bit tasteless compared to its peer in Cantonese cuisine. In fact, this dish is common on local households' dining tables, especially when they eat noodles. It is supposed to be light in flavor. After all, a simple life is a happy life.

真
风味斋

04

东城区锡拉胡同 14 号 · 14 Xila Hutong, Dongcheng District

去锡拉胡同，尝一回满满肉馅的“Beijing Pie”（“北京肉饼”）
Taste Meaty Beijing Pie in Xila Hutong

在王府井附近，您还能找到人均50块钱吃到“肚歪”并且余香满口的小店，一定要多给几颗星鼓励珍惜。位于锡拉胡同的河沿肉饼Beijing Pie——作为一个老派食客，一看到这个中英文店名开始犹豫了下：不会是调和中西方口味的改良派吧？结果呢，人家就是地道的京味儿肉饼。

他家的牛肉饼基本上每桌必点，皮儿薄馅儿大、满口留香，再沾点儿醋解腻，配个小米儿粥，满足了我的老派中国胃。门脸儿不大，但建议提前排号或者错过最高峰去——无论刮风下雨都得“等位”，哪怕是游客少的淡季，周边的本地居民都照样得等位半小时起步，人气就是这么旺。

It isn't easy to find a restaurant where one can enjoy a nice meal at the cost of only about 50 yuan in the bustling Wangfujing area in Beijing. The Heyan Roubing Beijing Pie in Xila Hutong is one of the few restaurants like this. An old-fashioned diner may hesitate at the moment when he sees its name combining Chinese and English, and wonder if it is a restaurant offering Chinese-foreign combinations. Actually, what it offers is authentic Beijing-style meat pie.

Beef pie is a must-try for almost every diner visiting here. Featuring a thin crust and rich filling, it is absolutely tasty. If you think it is a bit greasy, you can add some vinegar and eat the beef pie together with a bowl of millet porridge. Such a combination fits my old-style taste. The small restaurant is always packed with diners, and you'll have to wait for a seat regardless of the weather. You're suggested making a reservation in advance or avoid the peak hour. Even in the off-season when the number of tourists plummets, the restaurant is still crowded with local customers, and one still needs to wait at least half an hour for a seat.

河沿肉饼
Beijing Pie

PUNK ROCK
NOODLE
吃面
PUNK ROCK NOODLE
TROOPER

东城区东公街 25 号教师研修中心南侧
South of the Teacher Training Center, 25 Donggong Street, Dongcheng District

去“鼓楼吃面”，来一盘摇滚味儿的“肉啃肉”

Experience “Rock and Roll” and Meaty Foods at Gulou Chimian

您先瞧这店名：鼓楼吃面·肉啃肉——一个敢把“肉啃肉”挂成店名的馆子，菜得多硬多豪横啊。这其实是一家具有浓郁音乐气质、中西融合的店，创始人是朋克乐队蜜三刀乐队主唱雷骏，“肉啃肉”就是“摇滚（rock and roll）”的英文谐音。

这倒真是应了“不想当摇滚主唱的艺术家不是好厨子”的说法。说回他家的菜，他们家最初的招牌菜是金牌蘑菇肉酱面——牛肉馅口蘑炸酱，中西合璧。除此以外就是“肉啃肉”：基本上一次满足肉食爱好者的各种需求和胃口。这里可以作为肉食爱好者的宝藏小店收藏一下，硬气啊！

Gulou Chimian Rou Ken Rou, also known as the Punk Rock Noodle, is a restaurant noted for meaty foods. It boasts rich musical elements and fuses Chinese and Western styles. Its founder, Lei Jun, is the lead singer of the Misandao punk band. “Rou Ken Rou” (literally, “Meat Eats Meat”) is homophonic to “rock and roll.”

Someone jokes that “an artist who doesn’t want to be a rock and roll lead singer isn’t a good cook.” The signature dish of the restaurant is mushroom and meat noodles, which feature fried beef and mushroom sauce and combine Chinese and Western culinary styles. Another signature dish is Rou Ken Rou, a combination of assorted meat that can satisfy the demand of all meat gourmets. The restaurant is absolutely an ideal choice for meat lovers.

06

通州区月亮河河滨路 18 号 · 18 Hebin Road, Moon River, Tongzhou District

去合兴楼，尝一回“御膳肘子王”
Taste King's Pork Hock at Hexinglou

此“月亮河”不是赫本老电影里的“moon river”（月亮河）——它属于副中心大运河的支流，这里有月亮河温泉度假村。如果您来副中心逛运河，可以捎带手去吃一回合兴楼·御膳肘子王。

服务员上菜也有讲儿，要“敲铜锣，唱三祝”，图个吉利。肘子也好，皮酥肉嫩咸香适中，小料儿配得特别齐，各种口味总有一款适合你。人生有很多治愈方式，有人去终南山下穿个大褂儿站桩修行得到“灵魂放空”，有人直接到合兴楼来一份刚出锅的薄饼卷肘子就可以得到“灵魂放空”——咬下去的那一刹那，大脑突然空白了，物我合一了。“啊？我刚才要说啥烦心事来着？对不起，肘子一下肚儿，忘了。”

The Moon River in Tongzhou District isn't the "Moon River" sung by Audrey Hepburn in *Breakfast at Tiffany's*, but a tributary of the Grand Canal. By the river is the Moon River Hot Spring Resort. If you travel the Grand Canal, you can taste King's Pork Hock, a dish in the style of the Imperial Kitchen, at the nearby Hexinglou restaurant.

When the dishes are served, waiters ring the gongs and sing to greet the guests. Its signature dish, King's Pork Hock, tastes crispy outside and mildly salty, which is served alongside various kinds of condiments, of which at least one can whet your appetite. There are many ways to find inner peace. Some choose to practice Taoism in Zhongnan Mountain, and others choose to "empty the mind" by tasting just cooked pork hock wrapped by a pancake. A bite is enough to make you forget all your troubles.

御膳肘子王
合興樓

07

怀柔区怀黄路与红西路交叉口北行 100 米路西 · 100 meters west of the intersection between Huaihuang Road and Hongxi Road, Huairou District

去群生大院，尝一回 1910 年的老字号“东坝驴肉”

Enjoy Dongba Donkey Meat in the Time-Honored Restaurant the Qunsheng Compound

群生大院稍微有点儿远，您要是去怀柔自驾游，或者去红螺寺回来，可以方向盘一扭：直接奔“群生大院”打个卡。群生大院是老北京东坝驴肉的老店。说说“东坝驴肉”怎么成了京城一绝：当年群生世家的旧址，在东坝“驴市街”，过去东坝大集时，是骡、马、牛、驴、猪的交易市场，官称“大牙行”——其中驴最多，酱驴肉作坊也就随之而生。

来这儿必点老汤驴肉和驴肉火烧。店主人现在是第五代传人，坚守“驴脾气”——物价涨不减物力，老手艺绝不添加。当然了，也都提醒一句，尽量错过小长假、暑假高峰去体验更好，什么神仙馆子也架不住人太多，是吧？

Qunsheng Compound is located in Huairou District in suburban Beijing. If you finish your self-driving tour of the famous Hongluo Temple in the district, you are suggested visiting the restaurant on your way back to taste Dongba donkey meat. The Qunsheng Compound is a restaurant specializing in donkey meat. How did Dongba donkey meat gain fame in Beijing? In the very beginning, the restaurant was located on Dongba's Donkey Market Street. There was a livestock market where mules, horses, cattle, donkeys, and pigs were traded at that time. The market named Dayahang was particularly noted for donkey transactions. As time went on, restaurants and shops specializing in donkey meat emerged in the locale.

Must-try dishes in the restaurant also include donkey meat in original soup and donkey meat sandwich. At present, the restaurant is run by its fifth-generation inheritor, who still maintains the traditional cooking techniques handed down from generation to generation. Of course, I have to remind you that the restaurant is always crowded on holidays, and avoiding the peak season allows you to have a better dining experience.

老北京醬驢肉作坊
百分百真驴肉

这年头，吃素比吃肉金贵多了。如今在北京请客吃饭，吃素才是最体面的打开方式之一。无论您是为了体面、为了养生还是天生就不爱吃肉的“胎里素”，在京城挑素菜馆子，绝对是一项修行。

Today, vegetarian diets are more popular than meaty ones, and it is decent way to treat your guests in a vegetarian restaurant. Whether for decency, health or habit, it requires a certain knack to single out a satisfying vegetarian restaurant in Beijing.

第八章
Chapter 8

爷今儿吃素

Going Vegetarian for One Day

东城区安定门街道五道营胡同 2 号
2 Wudaoying Hutong, Andingmen Sub-district, Dongcheng District

去京兆尹，吃一顿纯素的“米其林三星”
Enjoy Vegetarian Foods at the King's Joy, a Three-Starred Restaurant in the MICHELIN Guide

这年头，吃素比吃肉金贵多了。如今在北京请客吃饭，吃素才是最体面的打开方式之一——比如去京兆尹，吃一顿纯素的“米其林三星”暨中国首家米其林“绿色之星”。

在北京，越是贵气的餐厅，门脸儿往往越低调，藏在灰秃秃的胡同里不显山不露水（有些甚至门口连招牌字样都没有），打开院门却发现别有洞天。“京兆尹”一词，原指的是古代官职。在北京的素餐里，京兆尹除了付款的那一下“肉疼”，几乎没有缺点：除了很多高端餐厅都能做到的食材菜单根据节令变化、重要食材全部原产地空运保鲜，以及米其林上餐顺序、摆盘大法乃至细节到一张餐巾纸的使用体验之外，京兆尹作为米其林素餐，最大的特点是不会特意把素菜做出肉菜的味道来满足肉食者的口腹之欲，吃的就是以最自然的方式，做出食材本身的味道——接近“淡然无味天人粮”的中国美学。当然了，如果您是习惯了高油高盐“重口”的食客，可能会觉得来这儿有点儿“花钱找罪受”：人均1600元，吃了一顿“不好意思说太淡了”的纯素餐。

Due to the popularity of vegetarian diets today, it is ideal to treat your guests in a vegetarian restaurant. As a vegetarian restaurant rated as three stars in the MICHELIN Guide and the first Chinese restaurant earning a "MICHELIN Green Star," the King's Joy is a good choice.

Many famous restaurants in Beijing do not have fancy appearances hidden in nondescript *hutongs*. Some even lack a noticeable signboard. Only when you get through the gates will you surprisingly find how special they are inside. The Chinese name of the King's Joy, *Jingzhaoyin*, refers to an official position in ancient China. Except for their high prices, the dishes offered here are perfect. Just like many other high-end restaurants, the King's Joy uses fresh ingredients transported from their origins by air, and meets MICHELIN standards in serving sequence, plate presentation, and even napkin folding. The most prominent feature of the King's Joy lies in that it doesn't deliberately make vegetables taste like meat to cater the appetite of meat gourmets, but preserves the natural flavor of ingredients. This style is close to the Chinese aesthetic concept that "natural taste is the best taste." Of course, those who are accustomed of heavy flavors may feel they are tasteless. Though all dishes it offers are vegetarian, the per capita spending here is about 1,600 yuan.

京兆尹
MICHELIN
KINGSJOY

02

东城区王府井大街王府中环西座 2 层
F/2, West Building, WF Central, Wangfujing Street, Dongcheng District

去山河万朵，尝一回“素食天花板”的艺术家仙气
Taste Artistic, Immortal Vegetarian Foods at the Vege Wonder

如果说京兆尹是贵气，那山河万朵就是仙气——不分伯仲。在“拒绝仿荤”这点上，山河万朵做得很彻底，也更接近“究竟”：您要是觉得素菜的最高水平就是“做出肉味儿”，那真是落伍了。山河万朵重新定义了人们对素菜的认知：素的最高境界，以中国的人文地脉为料理轴线，运用大自然的蔬果做出自然本味的中国素食——每一道料理背后的理念，是“山川河流”，是自然为道，咬下去一口，像是庄子的地籁与天籁在口腔里一起歌唱。

在这里，摆盘格调是绝对的艺术家气质——直接改改就能当美院作品展的水平。你看到这摆盘大法里的向日葵了吗？对对，相信你的眼睛，它就是一朵向日葵。谜底就在谜面上。来，致敬高级。

If the King's Joy boasts an atmosphere of dignity, its counterpart the Vege Wonder features an immortal ambience. In the pursuit of preserving the natural flavor of vegetarian ingredients, the Vege Wonder even works harder. You fall behind the times if you still think fine vegetarian foods should taste like meat. The Vege Wonder has redefined vegetarian foods. It uses Chinese culture and geography as the inspirations, creating a vegetarian culinary style unique to China. Behind each dish are "rivers and mountains" in nature. Taking a single bite, one may feel as if the sounds of nature reverberated in your mouth.

Its plate presentation is artistic, making every dish look like an artwork. When something looks like a sunflower in a plate is served, trust your eyes—it is truly a sunflower.

山河万朵
VEGE WONDER

03

朝阳区光华路2号6号楼阳光100国际公寓F座
Tower F, Sunshine 100 International Apartment, Building 6, 2 Guanghua Road, Chaoyang District

去叶叶菩提，尝尝二十四节气里的“百味养生”
Experience Herbal Foods Based on the 24 Solar Terms at the Yeye Bodhi

叶叶菩提也是京城里知名的素菜馆。您先看这店址：在CBD（中央商务区）、王府井这些“人民币味儿与加班脱发同在”的地段开素菜馆火出圈的，格调都低不了。他家很多招牌菜品，都是根据二十四节气调的养生菜，并且专门有研发小组，主打的就是一个“养”字：“养生”“养命”“养健康”。

比起“京兆尹”和“山河万朵”，“叶叶菩提”的人均消费要低点儿，但依然属于中高端意境素菜——人均700元左右。他家佛跳墙做得极好，基本上每桌必点。

The Yeye Bodhi is a famous vegetarian restaurant in Beijing. Popular vegetarian restaurants in CBD, Wangfujing, and other bustling areas are often expensive and stylish. Many signature dishes offered by the Yeye Bodhi are herbal foods invented on the basis of the 24 solar terms by the restaurant's R&D team. The key purpose is to preserve health.

Compared to the King's Joy and the Vege Wonder, the Yeye Bodhi is a bit cheaper—about 700 yuan per person. The restaurant is particularly known for Fotiaoqiang, which is a must-try for diners.

04

东城区和平西街化工大院社区内 · Huagong Courtyard Community, Heping West Street, Dongcheng District

去静莲斋，体验一回平民消费的“素菜荤吃”
Enjoy Affordable Vegetarian Foods at Jinglianzhai

哎呀，终于找到一个日常消费不心疼的素菜馆！比起上述“仙儿气”十足的素菜馆，静莲斋是平民口味的大众消费，朴素的素菜荤吃，素宫保鸡丁、素二指禅、素烤鱼……口味家常，比起轻食来说，毕竟是炒菜为主，略微油盐大了一点儿，但在可接受范围内——比较下饭。特别适合日常都是“白水煮青菜、几乎不放油盐”的严苛素食主义者，来静莲斋点个“素宫保鸡丁”，就算是给味蕾“解馋”了。

距离雍和宫不到两公里，对于有忌口的朋友来说是个福音。另外，低油低盐的纯素确实健康，但是过分严苛地“纯素+少油少盐”，容易心情抑郁。所以，还是得来静莲斋，沾一点儿素菜界的“不舍人间气”。

Jinglianzhai is a vegetarian restaurant suited for ordinary customers. Unlike the above-mentioned high-end vegetarian restaurants, it offers affordable, simple dishes that taste like meat, such as vegetarian Kung Pao chicken and vegetarian grilled fish. Compared to light meals, its dishes contain more oil and salt, but remain within an acceptable scope. Besides, they are more appetizing. Veggies who have been tired of tasteless boiled vegetables can eat vegetarian Kung Pao chicken here to satisfy their craving.

The restaurant is less than two kilometers away from the Lama Temple, which provides an ideal place to eat for Buddhism practitioners. Besides, though vegetarian dishes with little oil and salt are healthy, overemphasizing health will make the food tasteless and dinners depressed. Jinglianzhai is one of the few vegetarian restaurants for veggies who love heavily flavored food.

静蓮齋

05

西城区前门西大街正阳市场 3 号楼一层 1-3（老舍茶馆旁）· No. 1-3, F/1, Building 3, Zhengyang Market (next to the Laoshe Teahouse), Qianmen West Street, Xicheng District

去素虎，感受一回“菜虎子”心心念念的素食自助

Enjoy Vegetarian Buffet at the Vege Tiger Restaurant

北京话有个词儿叫“菜虎子”——专门形容特别能吃菜的人。素虎是个前门附近的素餐厅，绝对管饱——人均不到70块钱的素食自助，有大概小一百种素食。有绿油油的素菜，也有仿荤素菜：素糖醋里脊、素糯米鸡、素宫保鸡猴头菇……自助区有点心、各色素饺子。家常味道，对于过年过节大鱼大肉“杀疯了”的朋友来说，改口吃一回养生素，或者陪吃素的家人朋友来聚个餐，都是不错的选择。

In Beijing dialect, *Caihuzi* (literally "vegetarian tiger") refers to those who love vegetables. A vegetarian restaurant near Qianmen, the Vege Tiger Restaurant is celebrated for its buffet that offers nearly 100 kinds of food. In addition to green vegetables, there are also vegetarian dishes that taste like meat, such as vegetarian sweet and sour fillet, vegetarian chicken with glutinous rice and vegetarian Kung Pao chicken with mushroom. The buffet also provides a rich variety of pastry, and vegetarian dumplings. The restaurant is an ideal choice for those who want to experience vegetarian foods after eating too much fish and meat during the Spring Festival and other festivals or those who need to treat vegetarian friends.

素虎
Restaurant
Vege Tiger Restaurant
净素自助餐

东城区东直门内大街 144 号 · 144 Dongzhimennei Street, Dongcheng District

去花开素食，尝尝人均消费不到二百块的米其林餐厅！
Dine at the Blossom Vegetarian, a MICHELIN Restaurant with a Per Capita Spending of Less than 200 Yuan

快来珍惜一下，作为一家米其林上榜的纯素餐厅，花开素食的人均消费不到二百块，这让小年轻们流下了激动的泪水。

花开素食的素菜风格属于中西混搭，既有传统的中式素炖、吉祥三宝、雪媚娘、版纳素酸汤鱼，也有西式的惠灵顿牛排、法式坚果焗香薯。有选择困难症的客人可以直接选“双人餐”，店家都给您配好了。花开素食的口感，属于拯救“陪朋友吃素的吃荤者”的救星：里边的“一指禅”烤肠，咬在嘴里就是烤肠的味道，但是没用任何荤腥料；小素肉可以做出小酥肉的感觉。怎么样？一个餐厅，吃素的和吃肉的都满意的口感，拿走推荐不谢。

Young veggies with a humble salary would feel excited when they are informed that the per capita spending at the Blossom Vegetarian rated in the MICHELIN Guide is less than 200 yuan.

The Blossom Vegetarian offers both Chinese and Western cuisines. There are traditional Chinese vegetarian dishes such as stewed vegetables and Xishuangbanna vegetarian fish in sour soup, and Western dishes such as beef wellington and french roasted sweet potatoes with nuts. There are also set meals for those who have choice difficulty. The restaurant is considered a “savor” by those who are forced to eat vegetarian food with their friends. Its vegetarian sausage tastes just like real roasted sausage, but doesn't contain any meat, so does its vegetarian crispy meat. The restaurant satisfies the appetite of both veggies and meat gourmets.

BLOSSOM
VEGETARIAN
花開素食

07

东城区锡拉胡同 20 号 · 20 Xila Hutong, Dongcheng District

去福慧慈缘素食餐厅，胃安静了，心就放松了
Satisfy Your Stomach with Mindful Eating at the Fuhuiciyuan Vegetarian Restaurant

离着王府井不到一公里的锡拉胡同，居然有家“大隐隐于胡同”的素菜馆和艺术品博物馆——福慧慈缘素食餐厅。进去后整个人都很放松，环境布置非常空灵，工作人员的态度也亲切友善——能让你看到发自内心的特别善意美好的微笑。

菜品价格不贵，素菜不奢华但非常用心：人家“松鼠鳜鱼”里的“刺”恨不得都用素食仿真做出来。如果在世俗的喧嚣中累了，难过了，想歇歇了，来吃顿素菜吧。胃安静了，心就放松了。

Hidden in Xila Hutong, less than one kilometer from Wangfujing Commercial Street, is a vegetarian restaurant and art museum named Fuhuiciyuan. As soon as you enter the restaurant, you will feel relaxed upon seeing its Buddhist-style decorations, and the staff will greet you with friendly, sincere smiles.

Though carefully prepared, the dishes it offers aren't expensive. For example, the chef here even imitates the fish bone when cooking sweet and sour vegetarian Mandarin fish. Those who get tired of the hustle and bustle of life can relax themselves in the restaurant by satisfying their stomachs with mindful eating.

福慧慈缘

08

朝阳区东草园胡同铂宫国际中心 B 座一层 0106
No. 0106, F/1, Tower B, Bogong International Center, Dongcaoyuan Hutong, Chaoyang District

去东草园胡同，感受一次“藏茶素火锅”
Experience the Tibetan Tea Vegan Hotpot in Dongcaoyuan Hutong

如果您吃腻了传统的素菜炒菜，又受不了“喂兔子”的轻食餐，想改个口儿，找一家胃和心灵都能得到安抚的素火锅，可以试试东草园胡同的一叶一菩提藏茶素火锅。他家的锅底是藏茶，可以直接喝，开胃解腻。“爱情扎萨”很有地域风情，黑木耳糕、霸王花、素排骨东坡……素食者表示太友好了！再也不用陪朋友在涮肉火锅店小心翼翼地点个素菜，然后怎么涮都是一个白水煮菜味儿。

藏茶素火锅唯一的“缺点”是食材太新鲜、香气太盛，完全打乱了“肉食动物”的节奏，原来素火锅这么好吃！——吃完了容易再也不想吃肉了，给我根金箍棒，我就可以护送师父西天取经了。

If you're tired of traditional vegetarian dishes and cannot withstand light meals, you should visit the Tibetan Tea Vegan Hotpot, a restaurant that can satisfy both your stomach and soul. The restaurant's hotpot uses Tibetan tea as the soup, which can be eaten directly and has the effect of whetting the appetite and cleansing the palate. Its signature offer, Love Lhasa, is rich in Tibetan flavor. Other dishes include black fungus cake and vegetarian Dongpo ribs. Here, veggies no longer need to worry about how to order vegetables when enjoying hotpot with friends.

The Tibetan Tea Vegan Hotpot boasts fresh ingredients, and its vegetarian foods are so flavored that they can make meat lovers fall into love with vegetarian hotpot.

葉一世界

09

海淀区苏州街甲 16 号（临近中关村图书大厦、农业银行）· A16 Suzhou Street (adjacent to the Zhongguancun Book Building and the Agricultural Bank of China outlet), Haidian District

路过北大，体验一回“草木禅房”的养生素食
Experience Healthy Vegetarian Diets at Caomu Chanfang near Peking University

文化人请客或二三知己小聚，除非是特别熟的朋友，否则总不好意思头一回见面就请人撸胳膊挽袖子吃炙子烤肉，闹哄哄的对方说啥都听不清。如果您在北大或人大附近，想找个环境清雅一点儿、方便喝茶聊天吃饭的地方，可以来草木禅房养生素食——距离北大（燕园校区）不到 1 公里，距离人大 1.5 公里，木质禅房的环境，花木扶疏，青烟袅袅，每张桌子都有帘子轻轻隔开，餐具讲究，隐私保护得比较好，服务员轻言细语，店里还有自制祛寒湿的艾绒垫——从内到外，把“健康素”做到了极致。

除了正餐还有下午茶，如果有“点菜困难症”，可以直接选他家的套餐——全部是配好的，上每道菜的时间和先后顺序都拿捏得很好，正餐的人均消费 500 元左右，可以试试。

When treating guests, intellectuals prefer quiet restaurants rather than noisy rotisseries unless they want to have a party with intimate friends. If you want to find an elegant place near Peking University or the Renmin University of China to drink tea or eat, Caomu Chanfang (literally, Thatched Buddhist Abode) is an ideal choice. Less than a kilometer from Peking University (Yanyuan Campus) and 1.5 kilometers from the Renmin University of China, the restaurant features wooden-structure rooms hidden in sparsely distributed trees, with the smoke of incense burners curling up. Tables with elegant dinnerware are separated by a cloth curtain for privacy, and waiters speak quietly. The restaurant also offers homemade mugwort floss cushions that can dispel cold and dampness. It puts into practice the concept of health in and out.

In addition to main meals, it also offers afternoon tea. For those who have “decidophobia,” they can directly choose the set meals—all dishes will be served in a proper sequence. The per capita spending here is about 500 yuan, but it is worth visiting.

人生五味，各有所爱。如果不能逃离平庸的生活，至少可以逃离平庸的菜汤。

这一章的“好这口儿”，精选的都是味蕾上略带刺激的餐厅：或辣或酸或甜，甚至还有药膳的苦。总之，总有一款适合您的心情。

Each person has his or her own favorite taste. If you cannot get rid of plainness in life, at least you can get rid of plainness in eating.

This chapter focuses on restaurants that offer appetizing flavors—spicy, sour, sweet, or even bitter. There is always one flavor you may like.

第九章
Chapter 9

好这口儿

Favor This Flavor

朝阳区新东路首开铂郡 5 号楼 1-022 · No. 1-022, Building 5, Shoukai Bojun, Xindong Road, Chaoyang District

去三里屯“姆们”，咬一口“人生苦短，姆们很甜”
Sweeten Your Life at the M Sweetie Cake in Sanlitun

姆们是家网红蛋糕店。店名“姆们”，其实是北京话“我们”的发音，被老板做成了店名。三里屯很艺术，蛋糕造型很有艺术感——网红的魅影红丝绒蛋糕和拿铁丝滑到怀疑人生。海盐奥利奥、勃朗峰栗子蛋糕……爱吃甜品的朋友可以来试试。

去店里吃蛋糕，普及一个“假装内行”的小知识：千万别咬文嚼字读“姆们”，那就贻笑大方了。您要问老北京，人家可能一愣才反应过来。更地道的发音是“mme”，一开始有“乌”的口型（“我”字的发音被吃掉了）。还是搞一块“莫吉托”蛋糕吧，奶油里加了青橘皮碎，奶味厚实而不腻胃。走着。

The M Sweetie Cake is an Internet-famous bakery. Its Chinese name *mumen* is actually inspired by the pronunciation for “us” (*mme*) in Beijing dialect. The cakes it offers are immersed with artistic touch. In particular, red velvet cake and latte are famous for their silky softness. Sweets fans can also try sea salt oreos, Monte Bianco with chestnuts, and other offers of the bakery.

Remember that the bakery’s name should be pronounced *mme* in Beijing dialect. Let’s go to taste a piece of Mojito cake made of cream and minced tangerine peel, which features a strong but not greasy milk flavor.

姆们
M Sweetie Cake

02

朝阳区三里屯路 11 号太古里北区 N4 三层（Kenzo 旁扶梯上三层）
F/3, N4 Taikooli North Zone (close to the Kenzo store), 11 Sanlitun Road, Chaoyang District

去元古云境，尝一口“颜值甜点”茉莉清茶酪
Taste Jasmine Tea Yogurt at Yuangu Yunjing

元古云境地处三里屯，属于稍微晚来一会儿就要等位的火爆创意菜馆。中餐西做，禅意十足，谷饲牛腹肉、黑松露炒野山菌都是招牌。餐厅的环境属于中式浪漫，手机随手一拍都是格调满满，服务非常有边界感和分寸感，对于轻度社恐或者注重隐私的客人来说，非常舒适。

他家比正餐更出名的是“颜值甜点”，写着书法纸条——茉莉清茶酪对应“春”，咬一口下去，茶香馥郁后味甘苦，点缀着隐约的茉莉花香，细腻清爽。不知不觉就蒯（kuǎi，北京话，用勺子挖着吃的意思）完了一个……嗯，忘了我的小码裙子和它的卡路里。

Yuangu Yunjing is a popular restaurant offering creative dishes fusing Chinese and Western styles in Beijing's Sanlitun area. Customers usually need to wait for seats in peak hours. The restaurant features a Buddhist-style interior decoration, and its signature dishes include flank steak and fried wild mushroom with black truffle. The environment sends forth a sense of Chinese romance, and one can easily snap an instagrammable selfie here. The restaurant offers thoughtful services while paying attention to diners' privacy, which is comfortable for those with social phobia.

The restaurant is even more famous for its desserts, each served alongside a corresponding piece of calligraphy. Jasmine tea yogurt, for example, corresponds with the Chinese character *chun* (meaning "Spring"). The yogurt features the slightly bitter flavor and fragrant smell of jasmine tea, which tastes soft and refreshing. It is so yummy that I ignore its high calories, although I need to lose weight to fit in a small-size skirt.

文古

03

朝阳区朝阳北路龙湖长楹天街东区四楼星巴克对面 · Opposite to the Starbucks, F/4, East Zone, Longfor Changying Paradise Walk, Chaoyang North Road, Chaoyang District

去雁舍，吃一回8岁到80岁都能吃的“辣菜”
Eat Spicy Foods Suited for Diners Aged between Eight and Eighty at Yanshe

其实传统的北京菜没有那么辣，老北京人的味蕾也没有那么嗜辣。不过近些年因为川湘辣味“霸榜全国味蕾”，年轻人好辣，北京的辣菜馆子也渐渐多了。推荐“雁舍”，一是因为他家的辣度接受度比较高，适合聚会——有特别地道无辣不欢的过瘾湘菜，也有不太能吃辣的朋友也喜欢的“改良菜”。如果您不想吃太重的口味，又不想吃相对清淡的淮扬菜系，雁舍可能是接受度较广的选择。

另外，雁舍的装修风格也和“扑鼻辣气没地儿躲”的传统湘菜馆不同，北欧风和新中式融合，餐厅木质调氛围，隔断既清新敞亮又照顾隐私，轻声细语适合朋友商务小聚——毕竟请喷了香水的美女吃顿辣，总不好弄得人家毛衣外套头发上都沾着热辣油的味儿，是吧？

In fact, traditional Beijing cuisine isn't spicy, and older-generation Beijingers don't favor spicy foods. However, with the popularity of spicy foods from Sichuan and Hunan among youngsters in recent years, restaurants offering spicy dishes have been increasing in Beijing, of which Yanshe is highly recommended. On the one hand, Yanshe offers not only authentic Hunan cuisine that is extremely spicy, but also reformed dishes that cater to diners who cannot withstand very hot foods. So, it is an ideal choice for those who like neither heavily-flavored dishes nor tasteless Huaiyang cuisine.

On the other hand, Yanshe's interior decoration blends Nordic and Chinese styles, making it different from traditional Hunan-style restaurants sated with the smell of chili. Wooden partitions not only make the restaurant spacious and refreshing, but also protect the privacy of customers. It is an ideal place to dine with your girlfriend. After all, it would be awkward if a well-dressed woman smelled like chili oil after a meal.

04

朝阳区建国门外大街 1 号国贸商城北区四层 NL4006
NL4006, F/4, North Zone, China World Mall, 1 Jianguomenwai Street, Chaoyang District

去麻六记，记得早晨 10 点排队“堵门”拿号
Queue Up to Wait for a Seat at Maluji as Soon as It Opens at 10 A.M.

说完了辣，咱们说说“麻”。麻六记的国贸店是永远在排队的网红店——尤其是落地窗边上风景好、能把中央电视台总部大楼尽收眼底的好位置，要么提前两周预定，要么上午10点钟一开门就赶紧去“堵门”排队取号。

麻六记，突出的就是川菜的“麻”香过瘾。好这口儿的朋友可以试试他家的招牌水煮鱼——舌头已经“报警”了，但胃还是“指使”你不由自主地一筷子接着一筷子停不下来。不不不，不是我要故意吃那么多，是……是水煮鱼、毛血旺、自贡酸菜鱼、辣子鸡先动的手！

Diners always need to wait for a seat at the Maluji restaurant in China World Mall, especially one close to the window through which diners can view the new CCTV headquarters. To get such a prime spot, one has to either book two weeks in advance or line up as early as 10 a.m. when the restaurant just opens.

Maluji features spicy Sichuan cuisine. Spicy food lovers can try its signature dish—fish filets in hot chili oil, which may numb your tongue. Even so, its yummy taste makes you reluctant to put down your chopsticks. "I don't want to eat so much; I just can't help it." One may argue. But, save some room for sautéed eel with duck blood curd, Zigong boiled fish with pickled cabbage and chili, spicy chicken, and many more.

麻六記
MALUJI

05

朝阳区北三环樱花西街 18 号贵州大厦 2 楼（近和平西桥）· F/2, Guizhou Hotel (near Heping West Bridge), 18 Yinghua West Street, North Third Ring Road, Chaoyang District

去贵州大厦，感受一回“驻京办”的地道酸汤鱼

Taste Authentic Guizhou-Style Fish in Sour Soup in the Guizhou Hotel

说完了辣和麻，我们接着说说酸。在北京会吃的食客都知道，一定不要错过“地利优势”——各省的“驻京办餐厅”都值得扫一遍，吃的就是一个“味儿正”：驻京办餐厅嘛，都是当地的厨子、当地的食材、当地的做法——不用担心杂糅迁就各地口味变成“改良失败四不像”。

如果您要是好酸汤鱼的那口“酸汤”，来贵州大厦餐厅肯定没错。菜单可以“闭眼点”——除了招牌酸汤鱼，还可以尝尝米豆腐、黑糖糍粑等贵州特色菜。另外，只要是能接待“散客”的驻京办餐厅，人均价格都在百姓可接受范围内，可以放心去。

True gourmets know that restaurants in Beijing liaison offices of local governments are definitely worth visiting, which offer authentic local cuisines. They usually hire native cooks and use endemic ingredients, so diners don't need to worry whether their dishes are authentic.

If you are a fan of fish in sour soup, you shouldn't miss the restaurant in the Guizhou Hotel, where the Beijing Liaison Office of Guizhou Provincial Government is located. Apart from the signature dish fish in sour soup, other Guizhou-style foods such as rice jelly and glutinous rice cake with brown sugar are also worth trying. Don't worry about the prices. In fact, all such restaurants offer affordable dishes.

贵州大厦
GUI ZHOU MANSION

06

朝阳区兴化路化工社区大院甲7号 · A7 Huagong Community Compound, Xinghua Road, Chaoyang District

去岐黄小馆，吃一回有发明专利的“养生药膳”

Experience Patented "Medicinal Diet" at Qihuang Xiaoguan

麻、辣、酸、甜都说完了，舌尖百味各有归经，这“苦”能做出好吃的菜吗？还真有一家。和平里附近有一家“岐黄小馆养生药膳”——听这名字，您就知道肯定跟药膳有关。《黄帝内经》也被称为“岐黄之学”嘛。药食同源，养生菜最难得的是既养生又好吃——他家的招牌佛跳墙申请了发明专利，调料只有盐，您尝到舌尖的“鲜味儿”全靠食材本身调鲜，绝没有粗制滥造的味精或添加剂味儿。当归牛肉、黄精红烧肉……道道菜有讲儿。心动了？走起啊。

另外提醒，他家的“姜母鸭”由于太火爆，必须至少提前一天预订。着急也没用——得46味中药慢慢炖。人均消费在400元左右，在北京的养生药膳中不算便宜，但也属于可接受范围，毕竟，一分药食一分工，不能减物力嘛。

Can bitter ingredients be used for cooking? The answer is yes. The "medicinal diet" offered by Qihuang Xiaoguan near Hepingli uses bitter medicinal herbs as the ingredients. *Qihuang* means "medicine" in Chinese, so the ancient traditional Chinese medicine classic *Yellow Emperor's Inner Canon* is also called "The Studies of *qihuang*." Traditional Chinese medicine emphasizes the homology of medicine and food. The most difficult part of cooking medicinal diet is how to make it both healthy and yummy. Fotiaoqiang, the signature dish of Qihuang Xiaoguan, has obtained the invention patent, which only uses salt as the seasoning, instead of various kinds of additives such as monosodium glutamate. Thus, it allows diners to enjoy the authentic flavor of fresh ingredients. Other famous dishes include stewed beef with angelica roots and braised pork with soy sauce and roots of Siberian Solomon's seal. Let's give them a try.

It is noteworthy that if you want to taste the popular ginger duck, you must make a reservation at least one day in advance because it takes more than 10 hours to stew the duck with 46 sorts of medicinal herbs. The per capita spending in the restaurant is about 400 yuan. It isn't cheap to have a medicinal meal in Beijing, but this price remains acceptable. After all, preparing medicinal diet costs a lot of manpower and precious ingredients.

岐黄小館

07

西城区西直门外大街德宝新园乙 15 号 · B15 Debao Xinyuan, Xizhimenwai Street, Xicheng District

去郭林家常菜，见一回“好久不见”的老朋友
Reunite with Old Friends at the Guolin Restaurant

如果你约朋友吃饭，都约到郭林了，第一证明你们交情够瓷实，第二证明大家都开始老了。如果你抱怨“以前北京的郭林遍地开花，现在怎么店越来越少了，还得重新搜导航”，那可能大家哑然一笑，暴露年龄了：自己真的“老了”。郭林更像是记录北京一个时代、一批人青春的家常菜馆。要说烤鸭，它有；宫保鸡丁、鱼香肉丝、水煮鱼，它也有。总之都是家常菜的家常味道，装修也停留在“中年人审美”的配色体系中。但作为一家知名的京菜馆，这里适合见“好久不见”的人，怀旧的酒配着怀旧的干炸丸子，也似乎没有太多话可说，也无需说太多的客套话，碰一杯：“好久不见”。

If you choose the Guolin Restaurant as the place to treat your guests, that means two things: one is you and your guests are truly intimate friends; the other is you're getting old. It is a signal of aging if you complain that there were so many branches of the Guolin Restaurant in Beijing once but now you had to depend on a navigation app to find one. Today, the Guolin Restaurant is more like an ordinary restaurant that evokes the middle-aged generation's reminiscence of their youth. It offers roast duck, shredded pork with garlic sauce, fish filets in hot chili oil, and all other common dishes. Even its interior decoration caters to the aesthetic taste of the middle-aged. As a famous Beijing-style restaurant chain, it is suited to reunite with friends after a long period of separation. A cup of wine and a plate of deep-fried meat balls arouse nostalgia for yesterdays. For true friends to reunite again, they don't need to exchange polite greetings, but drain the cup with one gulp. "Long time no see! How are you doing these years?"

08

东城区后永康胡同 16 号 · 16 Houyongkang Hutong, Dongcheng District

去后永康胡同，到“做书咖啡”点杯美式撸个猫

Sip a Cup of Americano while Playing with Cats at the Zuoshu Café

“读书无解，来杯拿铁”。如果您生活在北京，连“做书咖啡”都没听过，那真是白在京城读书出版圈混了。您耳熟能详的许多知名作者、编辑、译者……都是这儿的“熟脸儿”。

北新桥一带的胡同，算是北京保留得最好的胡同区之一。从雍和宫或北新桥地铁口出来，跟着导航，七绕八绕九绕十绕（开玩笑）……就找到了曲径通幽的后永康胡同，然后能看到一个极其低调的牌子：“做书”。有小酒，有咖啡，有屋顶露台，有猫（并且让撸的那种），有满架子更新的书。到这来，来杯“做书美式”，千万别问邻座“是干纸质书苦，还是干一杯美式苦?”“你瞧你没境界了吧? 小朋友才做选择题。成年人选‘都苦’。”

If you get tired when reading, why not have a cup of latte? You must be an outsider of Beijing's publishing community if you never hear about the Zuoshu Café. Many renowned authors, editors, and translators are frequent visitors of the coffee shop.

Beixinqiao boasts the best-preserved *hutongs* in Beijing. Getting off the Yonghegong or Beixinqiao subway station and navigating through the labyrinth of alleys, you'll reach Houyongkang Hutong, where you can see a café with a nondescript signboard carrying the words *zuoshu*. In addition to drinks like wine and coffee, there is a rooftop platform where you can play with cats while enjoying the sunshine. There are also many shelves filled with books. Sip a cup of Americano and pick a book to read. Remember, don't ask the one sitting next to you which is bitter—reading books or drinking coffee? A binary choice is only for children, and an adult may answer that both are bitter.

做書
ZUO SHU

在北京混，食客圈流传着一条不成文的说法：“老饕签办”。

“老饕签办”，指的是对美食有极高要求的“老饕”，已经不需要“几颗星”来装点一顿饭的门面。他们有着随意与刁钻的味蕾自信——地道又低调的“驻京办”餐厅，才是“会吃”的标配。

A widespread saying among foodies in Beijing goes, "True gourmets prefer restaurants in Beijing offices of local governments."

True gourmets don't need to flaunt their social status by eating in luxury restaurants; they would rather follow their own hearts when choosing a place to eat. In their eyes, simple but authentic foods offered by restaurants in Beijing offices of local governments are for real gourmets.

第十章
Chapter 10

我爱驻京办

Delicacies Hidden in Beijing Offices of Local Governments

西城区南宁大厦 4 层 · F/4, Nanning Hotel, Xicheng District

去南宁大厦嗦一碗老友粉，据说连建筑都腌入味了

Taste Old Friends Rice Noodles at the Guangxi-Style Restaurant in the Nanning Hotel

问："广西人初到北京，如何安全度过第一个又干又冷又思乡的冬季?"答案是建议非药物治疗——交替服用南宁大厦那兰酒楼的老友粉与螺蛳粉，轻度思乡症一碗即愈，重度思乡症建议多加一份鸭脚煲或啤酒鸭一起嚼服。

南宁老友粉里边的酸笋、柳州螺蛳粉里的螺蛳肉，其"霸道"程度类似于北京小吃里的豆汁儿——好这口儿的人嗜之如命，要越浓郁才越勾魂儿；受不了这味道的人弃之如敝屣，别说去吃，隔着八米闻见味儿就"望风而逃"。要尝一碗配方地道、浓郁醇厚的螺蛳粉，还是得去南宁大厦。无论天儿多冷，螺蛳粉的汤一直咕嘟咕嘟在冒着热气——每天能卖出去多少份儿呢?据说南宁驻京办连建筑都腌入味了。

What can help a visitor from southern China's Guangxi Zhuang Autonomous Region survive his or her first dry and cold winter in Beijing? A diet recipe may work. A bowl of Old Friends Rice Noodles or river snail rice noodles from the Nalan Restaurant in the Nanning Hotel, both snacks of Guangxi, can cure slight homesickness. For those who are seriously homesick, they are suggested eating duck feet stewed in Chinese clay pot or stewed duck with beer.

Just like *douzhir* in Beijing, both Nanning Old Friends Rice Noodles and Liuzhou river snail rice noodles are well-known for their strange flavors, which are considered unacceptable by many. But for those who like such foods, the stronger the flavor, the better they taste. Nanning Hotel is where one can taste an authentic bowl of river snail rice noodles. No matter how cold the weather is, the soup is always steaming, emitting an odorous smell. No one know how many bowls are sold a day. Someone joked that its stinky smell has immersed the entire building.

南寧大廈

02

西城区西中胡同 28 号宜宾招待所 · Yibin Guesthouse, 28 Xizhong Hutong, Xicheng District

去宜宾招待所点一回地道川菜，感觉前半生吃过的“鱼香肉丝”都白吃了

Enjoy Authentic Sichuan Cuisine at the Yibin Guesthouse in Beijing

问：去驻京办吃饭，如何不踩坑呢？有一个“老饕秘诀”——别贪大。市级的不一定就比省级的味道差，越偏越不起眼的驻京办甚至招待所，往往味道越“神秘”，倔强而正宗：因为它不需要调和一省之口味，它只需要代表自己。

真正嘴刁的人去川菜馆子，先点“鱼香肉丝”看大厨的水平——越是简单的菜越考验师傅的技术。一个川菜馆要是连鱼香肉丝都能做得“味正”，说明其他菜都差不到哪里去。宜宾招待所就属于连鱼香肉丝都能做得特别正的川菜馆——吃完感觉自己前半生吃的“鱼香肉丝”都是高仿版，白吃了。当然，来这儿不能只吃家常菜，这里点单率最高的招牌是李庄白肉：刀功没有个三五年都切不了这么薄的片，薄如蝉翼大致如此，用筷子卷起白肉蘸着小碗里的蒜蓉料，这口感真是绝了，肉不柴不腻入口即化的感觉，咸鲜肥美，蒜香浓郁。唯一的缺点就是排队的人太多了……基本上等位一小时起步。如果周末您去吃饭，发现居然不用等位，那真是激动的心、颤抖的手，足可以在饭桌上吹一年的牛。

Veteran gourmets suggest people eat in Beijing liaison offices of municipal governments rather than those of provincial governments because mysterious and authentic regional foods are often found in nondescript guesthouses opened by the municipal governments of various cities in Beijing. They don't need to make compromises to satisfy the different tastes of people from the whole province, but only need to cater to the unique taste of people from the same city.

Gastronomes can judge the skills of a Sichuan-style restaurant's chef according to the shredded pork with garlic sauce he cooks. The simpler the dish, the greater cooking skills it requires. If a restaurant can make authentic shredded pork with garlic sauce, it must be able to cook other dishes well. If you have a chance to taste shredded pork with garlic sauce in the Yibin Guesthouse, you'll find that the same dish you have eaten before seems like a "counterfeit."

Of course, you shouldn't come here only to taste regular dishes. The most popular signature dish is Lizhuang pork with garlic sauce. The dish requires superb cutting skills. The pork is cut into slices as thin as a cicada's wings. Dip a slice in the garlic sauce, and it tastes incredibly yummy as if it would melt in your mouth. The only problem is that you'll have to wait at least an hour for a seat. You must be extremely lucky if you don't need to line up for a seat here on weekends.

宜人宜賓

03

朝阳区西坝河南路光熙门北里 15 号 重庆饭店内
Chongqing Hotel, 15 Guangximen Beili, Xibahe South Road, Chaoyang District

去重庆饭店，辣子鸡的辣椒都可以打包的水准

Pack the Leftover Chili Peppers after Enjoying Spicy Chicken in the Chongqing Hotel

作为重庆驻京办的餐厅，重庆饭店是嗜辣一族的狂欢之地。他家的渝菜几乎是“闭眼点”的水平，火锅自不必说——重庆火锅在京城的热度比四川火锅还高。炒菜不踩雷，连一个简单的炝炒圆白菜都可以做得很好吃。点单率最高的除了招牌水煮鱼，还有辣子鸡——辣子鸡去骨留皮，去骨是为了方便放心大口嚼；略带皮是为了炸得更酥脆，一口咬下去油香四溢，再加上辣子鸡煸炒得很干很到位，油而不腻。整体口味偏麻（而非纯辣），辣椒的层次感处理得很丰富，不仅鸡块要风卷残云，盆底的辣椒也一定要打包——回家随便炒个蔬菜都非常“巴适”(四川方言，指很好、舒服、合适，亦指正宗、地道)。除了主菜，最后别忘了加一份红油抄手或担担面做主食。

The restaurant in the Chongqing Hotel, where the Beijing Office of Chongqing Municipal Government is located, is a paradise for spicy food lovers. Almost every Chongqing-style dish in the restaurant is mouthwatering, let alone its signature hotpot. In fact, Chongqing-style hotpot is even more popular than Sichuan-style hotpot in Beijing. Even regular dishes like stir-fried round cabbage are very tasty. In addition to fish fillets in hot chili oil, spicy chicken is one of the most welcomed offers. The chicken in the dish is boneless, so that diners don't need to worry about their teeth when chewing in gulps. The deep-fried skin tastes crispy but not greasy. Even the chili peppers are well-processed. Thus, after you finish eating the chicken, don't forget to pack the leftover chili peppers, which are suited to use as seasonings when you cook at home. After the main courses, you can order a bowl of spicy wontons in chili oil or noodles in chili sauce.

北京重庆飯店
BEIJING CHONGQING HOTEL

04

海淀区中关村南大街36号湖北大厦1楼
F/1, Hubei Hotel, 36 Zhongguancun South Street, Haidian District

去湖北大厦，来一份“神仙水”级别的莲藕排骨汤
Taste the Yummy Pork Chop Soup with Lotus Roots in the Hubei Hotel

如果您从国家图书馆刚刚享受完“精神食粮”出来，想给自己的皮囊也来顿“大餐”，可以果断放弃国图附近的“快餐”，直接走五六百米，到马路对面的湖北大厦一层的湖北味道，吃一顿地道的驻京办大餐。驻京办餐厅丰俭由人，您只点一份热干面也能做，配一份孝感米酒，马上“梦回武汉”。如果是二三知己一起前往，可以点招牌的武昌鱼、武汉三鲜豆皮、干锅鸡、粉蒸肉……样样不差。如果是秋冬季节，强烈推荐“神仙水”级别的莲藕排骨汤——和湖北煲的汤一比，我这个老北京感觉自己之前做的“莲藕汤”，在武汉只能算“刷锅水”。湖北的“莲藕排骨汤”，用的是粉藕，拉丝很长，口感软糯；排骨的火候很足，极为软烂。最重要的精华在汤，鲜甜咸醇，入口的层次与冲击感极强，粉藕的清甜融在骨汤的馥郁里，是极好的滋阴养生膳，感觉气血直接从嗓子喝进去，“补进”了五脏里。可以一试。

After nourishing your mind at the National Library of China, you may want to nourish your body with yummy foods. Ignore the fast-food restaurants next to the library, walk 500 to 600 meters to go directly to the Hubei Hotel on the opposite side of the street, where authentic Hubei cuisine won't let you down. The hotel houses the Beijing Office of Hubei Provincial Government as well as a Hubei-style restaurant offering a variety of dishes at different prices. You can order a bowl of Wuhan Hot Noodles with sesame paste, paired with a cup of Xiaogan rice wine, which may generate an illusion that you were in Wuhan, the capital of Hubei Province. If you come here with several friends, you can try signature dishes include steamed Wuchang fish, Three Delicacies Wrapped in Tofu Skin, sautéed chicken with pepper in iron wok, and steamed pork with rice, among many other Hubei-style delicacies.

In autumn and winter, I strongly recommend you taste pork chop soup with lotus roots here, which makes me feel that all the lotus root soup I've cooked myself is tasteless. To make this dish, Hubei cooks use starch-rich lotus roots which taste soft, and stew the pork chops for hours to make them pulpy. The essence of this dish lies in the salty but fresh soup. A little sip is enough to tickle your taste buds. The refreshing and sweet taste of lotus roots is mixed into the thick soup, which can nourish the Yin and improve your health. It is definitely a must-try.

朝阳区北四环路惠新西街 1 号 安徽大厦中餐厅 · Chinese-style Restaurant, Anhui Tower, 1 Huixin West Street, North Fourth Ring Road, Chaoyang District

去安徽大厦，点一份惊艳的臭鳜鱼
Taste the Amazing Smelly Mandarin Fish in the Anhui Tower

作为一个有南方胃的北京人，我爱徽菜。徽菜既满足了汤汤水水的润感，又不像淮扬菜那么清淡考究到吃起来容易有心理压力，也不像粤菜的甜口儿会使北方人的“味蕾水土不服”，还不用担心像两湖和川渝菜那样大半个菜单都是辣菜，对不吃辣的人也极度友好。安徽大厦属于广义的“徽菜”——既有皖北菜，也有江淮之间及皖南徽菜。但无论您是更偏皖南还是皖北口味，到了安徽大厦，臭鳜鱼基本上是每桌必点——蒜瓣儿肉，汤汁浓郁咸鲜，闻起来有一股“臭味”，但吃起来香。所谓臭味，其实是鳜鱼在腌制过程中微生物被慢慢地分解掉所散发出的独特的气味，经过如此秘制后的臭鳜鱼，还多了一层普通烹制所无法调和出的鲜香气，蘸着汤汁，越吃越上瘾，属于“下饭利器”。除了臭鳜鱼以外，毛豆腐、石锅老豆腐或者徽州老鸡汤，都是这里的常销招牌。哎，念及舌尖痴绝处，无梦常到徽州诉。可以一试。

Despite being a Beijing native, I have a stomach catering to cuisines of southern China, especially Anhui cuisine. Anhui cuisine satisfies my desire for yummy soups. Moreover, it is neither as tasteless and over-exquisite as Huaiyang cuisine, nor does it taste as sweet as Cantonese cuisine or as spicy as Hunan, Hubei, Sichuan, and Chongqing cuisines.

The restaurant in the Anhui Tower offers signature dishes from all over Anhui Province, covering regions on both banks of the Huaihe River that flows through the province. Smelly Mandarin fish is a must-try in the restaurant. Featuring delicious fish and refreshing soup, the dish doesn't smell good but tastes yummy. The odor comes from the microorganisms produced in the salting process of the Mandarin fish. However, the Mandarin fish processed in this way has a refreshing flavor that is unable to obtain in other cooking methods. It is appetizing especially when mixed with dipping sauce. In addition to smelly Mandarin fish, other popular signature dishes include stinky tofu braised in soy sauce, stewed tofu in stone pot, and Huizhou chicken soup. All of them are worth trying.

安徽大厦
AN HUI TOWER

06

东城区北京站街 9 号 湖南大厦 · Hunan Hotel, 9 Beijingzhan Street, Dongcheng District

去湖南大厦，数一数剁椒鱼头能下几碗米饭

Enjoy Steamed Fish Head with Diced Hot Red Peppers in the Hunan Hotel

湖南驻京办餐厅的地理位置太好了——北京火车站对面。一桌地道湘菜，抚慰了南来北往游子的心。如果赶上北京秋雨降温，又冷又冻没来暖气的初冬，最佳解决方式就是去湖南大厦吃一顿够辣又入味的剁椒鱼头，配一碗白米饭吃到额头冒汗，从胃里暖到脚底；再来一份小炒黄牛肉或者辣椒小炒肉，辣椒的香辣，肉的酱香，非常巧妙地融合在一起，不仅不油，而且辣椒比肉好吃，口感丰满，辣椒都能吃出回甘的滋味。最后再来一份清炒红菜薹爽口——除了食材的新鲜，餐厅的火候控制极好，一个普普通通的素菜都拿捏得恰到好处，感觉多两秒就老，少两秒就不透，蔬菜的清甜和油香平衡得特别好，不需要蒜来爆香都非常可口。

The restaurant in the Hunan Hotel, where the Beijing Office of Hunan Provincial Government is located, enjoys a prime location facing the Beijing Railway Station on the opposite side of the street. A bite of authentic foods from their hometown can help Hunan people drifting in Beijing relieve homesickness. Especially on a cold rainy day in late autumn and early winter, it is a great pleasure to taste the spicy steamed fish head with diced hot red peppers at the restaurant in the Hunan Hotel. Paired with a bowl of rice, the dish can warm up diners from head to toe.

Other recommended dishes include stir-fried beef with chili pepper and stir-fried pork with chili pepper. Spicy chili and the flavor of the meat blend to tickle your taste buds. In the two dishes, chili peppers even taste more palatable than meat, with a lasting aftertaste. At last, you can order a vegetarian dish such as sautéed field mustard to offset the flavor of spicy meat. In addition to fresh ingredients, the vegetarian dishes are perfectly cooked—neither more nor less. The fresh vegetable and the edible oil are well balanced, which tastes yummy even without flavorings such as garlic.

07

东城区东花市北里东区 7 号楼与 8 号楼中厅 1 层 · F/1, Central Hall between Building 7 and Building 8, East Zone, Donghuashi Beili, Dongcheng District

去云腾食府，从此爱上云南的米线和各种蘑菇

Fall in Love with Yunnan Rice Noodles and Mushrooms after a Visit to Yunteng Shifu

作为云南驻京办的老牌餐厅，云腾食府不在云南大厦，而在驻京办的云腾宾馆。之前有朋友抱怨，说在北京不太容易吃到地道的云南米线外卖——做一碗低调正宗、不按北方口味调和改良的米线，如果再加上底汤火候、地道食材，各环节用工用料不省，一碗米线合算起来并不便宜，再加上外卖的“锅气”减了一大半儿，基本上销量惨淡，竞争不过“改良版”的平价米线。要吃到带着锅气又正宗的云南过桥米线，还真值得跑一趟云南驻京办餐厅——底汤很鲜很香，汤头可以喝出鸡的味道，是真正的鸡汤熬煮的，不少朋友一碗不够，再添了一碗。汽锅鸡、竹筒米酒、黑三剁基本上是每桌必点的招牌，至于各种“蘑菇开会”的特色菜更毋庸多言，建议点时令或者还没尝过的口味，每次都开发一下人生新鲜感。当然，唯一的友情提示是，不熟悉性味的蘑菇要慎重点，如果赶上北京春季容易上火的时候，最好选择性味平和一些的蘑菇，或者要搭配其他菜品，万不可“万蘑齐发”，不然容易上火牙疼。

A time-honored restaurant under the Beijing Office of Yunnan Provincial Government, Yunteng Shifu isn't in the Yunnan Hotel, but in the Yunteng Hotel. A friend once complained that he couldn't find authentic Yunnan rice noodles in Beijing. Actually, compared to rice noodles that are reformed to cater to the taste of people in the north, it is costly to cook authentic Yunnan rice noodles, considering the intricate cooking techniques and endemic ingredients required. Besides, the taste of takeout is inferior to that of just cooked rice noodles. Therefore, authentic Yunnan rice noodles are relatively expensive, making it lack competitiveness in the face of various kinds of "reformed" rice noodles that are cheaper.

If you want to try authentic Yunnan rice noodles, Yunteng Shifu is worth visiting. The rice noodles it offers feature fresh, yummy soup cooked with chicken. Many diners will ask for another bowl. Other must-try offers include steam pot chicken, rice wine in bamboo tube, and Heisanduo (sautéed chopped pork), let alone the special signature dish Assorted Mushrooms. You can choose seasonal mushrooms or categories you have never tried before. One thing to know: be careful ordering unfamiliar mushrooms, for their medicinal effects may cause unpleasantness. For example, in spring when people are vulnerable to internal heat, you'd better choose the mushrooms with an effect in relieving pathogenic heat or pair them with vegetables. Otherwise, eating too many mushrooms may cause toothache.

08

朝阳区花园路街道马甸桥南路 2 号七省大院 江西驻京办内 · Beijing Office of Jiangxi Provincial Government, Qisheng Compound, 2 Madianqiao South Road, Chaoyang District

去赣人之家，报恩的瓦罐汤和“报仇”的辣
Taste Jiangxi-Style Clay Pot Soup at Ganren Zhijia

之前有江西朋友跟我开玩笑，如果要在中国找最低调省份，江西肯定能排进前三。但如果你认为江西菜的存在感也很低，那就大错特错了。来赣人之家，“瓦罐汤自选”是招牌，十几种瓦罐汤可以自己选，点单率最高的是乌鸡汤，汤色清澈，味道鲜甜不腻口，毫无科技感，真的是慢火真料熬出来的，不出北京就实现了“瓦罐汤自由”，不过因为每天定量熬制又太火爆，去晚了就没了。江西米糕有淡淡的发酵香味，气孔绵密，口感细腻绵软却有嚼劲，推荐一定要趁热吃，打包回家再加热，就不是一个味儿了。此外，井冈烟熏笋、南昌啤酒鸭……基本上实现了赣菜又咸又辣又香得停不下来的口感集合体，极下饭，一个人能干半盘子，第一口吃着觉得咸甜适中，辣度不显，结果越吃越辣，“嘶哈嘶哈”却停不下来。如果确实不是“吃辣选手”，服务员在问“正常做还是微辣”时，建议不要打肿脸充胖子，不然赣菜的威力……

A friend from Jiangxi Province once joked that Jiangxi would be ranked among the three lowest-profile provinces in China. You would be wrong if you thought Jiangxi cuisine is “unknown” too.

At Ganren Zhijia, a restaurant opened in the Beijing Office of Jiangxi Provincial Government, there are a dozen kinds of clay pot soup, a representative of Jiangxi cuisine, for diners to choose. The most popular is black chicken soup, which features clear soup that tastes fresh and sweet. The soup is stewed with fresh ingredients for hours. However, you'd better go there early because the soup may be sold out soon. The restaurant also offers Jiangxi rice cake, which features a light fermented smell and dense pores and tastes soft but chewy. It is suited to be eaten on the spot since its taste will deteriorate if reheated.

Other signature dishes include Jinggang smoked bamboo shoots and Nanchang stewed duck with beer. Jiangxi cuisine fuses various flavors—salty, spicy, and oily, which is highly appetizing. You may feel it isn't hot at the first bite, but the spicy taste is getting stronger and stronger. Thus, when the waiter asks what degree of spiciness you want, don't choose “highly spicy” unless you fear no spicy foods.

人之家

09

朝阳区河南大厦 2 层 · F/2, Henan Hotel, Chaoyang District

去河南大厦，干一碗魂牵梦绕的羊肉烩面
Taste a Long-Waited Bowl of Stewed Noodles with Mutton in the Henan Hotel

去河南大厦吃饭，一定慎重穿浅色的衣服，并且不推荐相亲来这里吃饭——因为你可能必点烩面，汤面太地道了，就容易尽情吃花了妆，还很容易“忘情”把汁水溅到浅色衣服上。回头算了算，粉底太贵，擦一回五十；衣服干洗一百，比一碗烩面还贵。很可能赚了一碗烩面，输了一个对象。

河南大厦的招牌菜是黄河鲤鱼、羊肉烩面、豫东小酥肉、胡辣汤和几乎每桌必点的主食烩面。羊肉烩面的灵魂不在羊肉，而在汤——一定要用略带肉的骨头配合大锅熬，火候要足，熬制至少五个小时，一直熬到汤白亮如牛乳，口感馥郁有章法，才算工成。黄河鲤鱼和小酥肉见

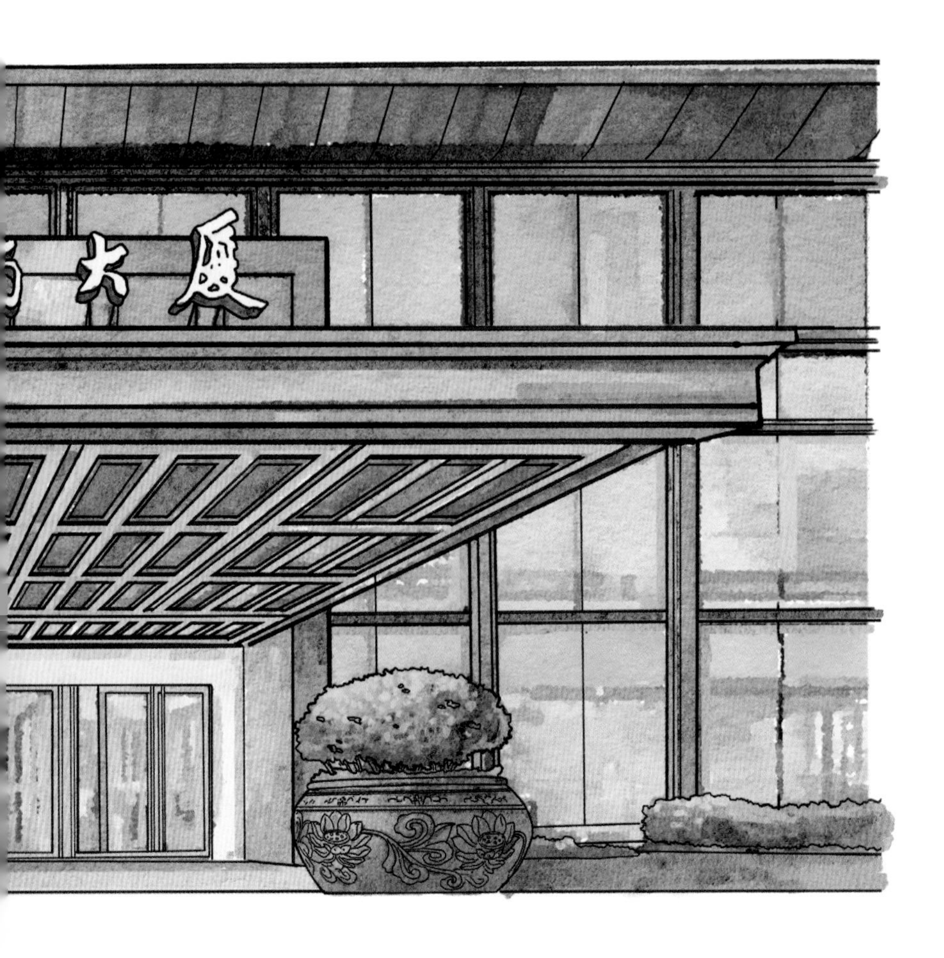

仁见智，喜欢的人爱它浓厚酸甜的宽汁儿；好咸鲜口儿的人可能觉得居然被做成了甜味儿，“必有妖孽”。说了这么多，你们看着点就行。

Don’t wear light-colored clothes when you dine at the restaurant in the Henan Hotel. Also, it isn’t an ideal place for first-time dating. All diners coming here may order its signature offer—stewed noodles with mutton. It is easy for girls to mess up their makeups and leave stains on their light-colored dresses when gobbling the yummy noodles. Perhaps reapplying makeup and cleaning stained clothes cost much more than the noodles. More importantly, you may leave a negative impression on your date.

In addition to the must-try stewed noodles with mutton, the signature dishes of the restaurant in Henan Hotel also include braised Yellow River carp, crispy meat of Eastern Henan, and spicy soup. The soul of stewed noodles with mutton isn’t the mutton but the soup. The soup is stewed with pork chops for at least five hours until it becomes as white as milk and emits a strong fragrant smell. Braised Yellow River carp and crispy meat feature sour and sweet soup. Some like them, but those who dislike sweet may consider their taste a bit unacceptable. It all depends on your taste.

10

朝阳区北三环安贞西里三区 26 号浙江食府 · Zhejiang Restaurant, 26 Anzhen Xili No. 3 Zone, North Third Ring Road, Chaoyang District

去浙江大厦，江南不仅有“西湖牛肉羹”
Taste West Lake Beef Soup and More in the Zhejiang Hotel

看徐克的老电影《青蛇》，西湖畔刚刚化作人形的白素贞约许仙吃饭，都要“尝一尝西湖牛肉羹”。白纱帷幔翩翩，半冷半暖秋天。我在小时候就在想，这道西湖牛肉羹是什么样的神仙味道。作为中国八大菜系之一的“浙菜”，说起来有点儿复杂——至少先粗分杭州菜、宁波菜、绍兴菜、温州菜等几大流派。如何不出北京就在一桌之内体验浙菜精华？可以去浙江驻京办的浙江食府：不仅有传说中的西湖牛肉羹，还有杭州特色的西湖醋鱼、龙井虾仁、东坡肉，宁波特色的宁式鳝丝，加入了金华火腿提鲜的笋干老鸭煲，无辣不欢口味偏重的朋友，还可以加一道衢州辣炒小公鸡，基本汇聚了浙江一省代表菜的精华。虽说比起浩瀚无边的浙江菜，驻京办的一桌餐只算是“一勺西湖水”，但一勺足以解忧。值得走一回。

In Hark Tsui's movie *Green Snake*, the White Snake treats her lover Xu Xian in her mansion on the shore of West Lake after she turns into a lady, and one of the dishes she cooks is West Lake beef soup. The white gauze curtains sway in autumn breeze, creating a dreamlike atmosphere. When I watched the movie as a child, I dreamed of tasting West Lake beef soup one day.

One of the Eight Great Cuisines of China, Zhejiang cuisine consists of many sub-genres such as Hangzhou cuisine, Ningbo cuisine, Shaoxing cuisine, and Wenzhou cuisine. Where can one taste authentic Zhejiang cuisine in Beijing? The Zhejiang Restaurant in the Beijing Office of Zhejiang Provincial Government is an ideal choice. Apart from West Lake beef soup, there are also Hangzhou-style dishes such as steamed grass carp in vinegar gravy, stir-fried prawn with Longjing tea, and Dongpo pork, as

well as Ningbo-style shredded eel and stewed duck with dried bamboo shoots and Jinhua ham. For spicy food lovers, fried cockerel with chili pepper, a famous dish of Quzhou cuisine, is recommended. All of the dishes represent the quintessence of Zhejiang cuisine. However, what the restaurant offers is merely a small portion of the plethora of Zhejiang-style dishes, but it is still worth visiting.